EXERTIONAL HEAT ILLNESSES

LAWRENCE E. ARMSTRONG, PhD

UNIVERSITY OF CONNECTICUT

EDITOR

HUMAN KINETICS

Library of Congress Cataloging-in-Publication Data

Exertional heat illnesses / Lawrence E. Armstrong, editor.
   p. ; cm.
Includes bibliographical references and index.
   ISBN 0-7360-3771-3 (hard cover)
   1. Heat stroke. 2. Heat exhaustion. 3. Syncope (Pathology)
   [DNLM: 1. Heat Stress Disorders. 2. Exercise--physiology. WD 610
E96 2003] I. Armstrong, Lawrence E., 1949-
   RC87.1 .E946 2003
   616.9'89--dc21
                                        2002152292
ISBN: 0-7360-3771-3

The Web addresses cited in this text were current as of January 29, 2003, unless otherwise noted.

Acquisitions Editor: Michael S. Bahrke, PhD; **Developmental Editor:** Renee Thomas Pyrtel; **Managing Editor:** Amanda S. Ewing; **Copyeditor:** Christine M. Drews; **Proofreader:** Sarah Wiseman; **Permission Manager:** Dalene Reeder; **Graphic Designer:** Andrew Tietz; **Graphic Artist:** Denise Lowry; **Photo Manager:** Leslie A. Woodrum; **Cover Designer:** Jack W. Davis; **Photographer (cover):** © Empics/ SportsChrome; **Photographer (interior):** © Human Kinetics, unless otherwise noted; **Art Manager:** Kelly Hendren; **Illustrator:** Accurate Art; **Printer:** Edwards Brothers

Printed in the United States of America 10 9 8 7 6 5 4 3 2 1

Human Kinetics
Web site: www.HumanKinetics.com

United States: Human Kinetics, P.O. Box 5076, Champaign, IL 61825-5076
800-747-4457
e-mail: humank@hkusa.com

Canada: Human Kinetics, 475 Devonshire Road Unit 100, Windsor, ON N8Y 2L5
800-465-7301 (in Canada only)
e-mail: orders@hkcanada.com

Europe: Human Kinetics, 107 Bradford Road, Stanningley,
Leeds LS28 6AT, United Kingdom
+44 (0) 113 255 5665
e-mail: hk@hkeurope.com

Australia: Human Kinetics, 57A Price Avenue, Lower Mitcham, South Australia 5062
08 8277 1555
e-mail: liahka@senet.com.au

New Zealand: Human Kinetics, P.O. Box 105-231, Auckland Central
09-523-3462
e-mail: hkp@ihug.co.nz

To the health and welfare of athletes, laborers, and soldiers who exercise or work in hot environments.

To the professionals who evaluate, diagnose, and treat exertional heat illness patients.

Lawrence E. Armstrong, Editor

CONTENTS

Foreword   viii
Preface   ix

Chapter 1   Physiology of Heat Stress   1
John W. Castellani

Heat Balance 1
Temperature Regulation and the Brain 2
Modifiers of Core Temperature Regulation 4
Control of Water and Ions 6
Physiological Responses to Exercise–Heat Stress 9
Factors That Limit Exercise Performance in Hot Environments 11
Factors That Modify the Exercise–Heat Stress Response 12
Summary 15
Disclaimer 15

Chapter 2   Classification, Nomenclature,
and Incidence of the Exertional Heat Illnesses   17
Lawrence E. Armstrong

Classification 19
Nomenclature 21
Incidence 23
Summary 28

Chapter 3   Exertional Heatstroke: A Medical Emergency   29
Douglas J. Casa, Lawrence E. Armstrong

Pathophysiology 31
Epidemiology and Prognosis 39
Identification of Exertional Heatstroke 40
Medical Treatment of Exertional Heatstroke 47
Considerations for Return to Training 51
Preventing Exertional Heatstroke by Understanding
the Most Common Scenarios 51
Summary 56

Chapter 4 Heat Exhaustion, Exercise-Associated Collapse, and Heat Syncope 57

Lawrence E. Armstrong, Jeffrey M. Anderson

Heat Exhaustion 57
Exercise-Associated Collapse 82
Heat Syncope 86
Summary 89

Chapter 5 Exertional Heat Cramps 91

Michael F. Bergeron

Etiology and Related Pathophysiology 93
Common Scenarios 94
Signs and Symptoms 95
Treatment 96
Continuation of Activity and Return to Play 98
Prevention 98
Summary 102

Chapter 6 Exertional Hyponatremia 103

Lawrence E. Armstrong

Common Scenarios and Predisposing Factors 104
Clinical and Physiological Features of Exertional Hyponatremia 105
Fluid–Electrolyte Status of Athletes With Exertional Hyponatremia 114
What Volume of Water Is Required to Induce
Exertional Hyponatremia? 121
Is Rhabdomyolysis Associated With Exertional Hyponatremia? 126
Treatment 129
Preventing Exertional Hyponatremia 133
Summary 135

Chapter 7 Minor Heat Illnesses 137

Robert W. Kenefick, Melissa P. Hazzard, Lawrence E. Armstrong

Transient Heat Fatigue 138
Anhidrotic Heat Exhaustion 140
Heat Edema 141
Miliaria 142
Sunburn 145
Summary 148

Chapter 8   Predisposing Factors for Exertional Heat Illnesses   151
Lawrence E. Armstrong, Douglas J. Casa

Exertional Heatstroke 151
Exertional Heat Exhaustion 158
Heat Syncope 159
Exertional Heat Cramps 161
Exertional Hyponatremia 162
Summary 166

Chapter 9   Considerations for the Medical Staff: Preventing, Identifying, and Treating Exertional Heat Illnesses   169
Douglas J. Casa, William O. Roberts

Responsibility of the Medical Staff 170
Reducing Incidence of Exertional Heat Illnesses 170
Identifying Hyperthermia 175
Defining Exertional Heat Illness 176
Treatment of Exertional Heat Illnesses 183
Staffing, Site, Equipment, and Supply Considerations 186
Recommendations for the Medical Team and Event Administration 192
Summary 195

Chapter 10   Recommendations for Athletes and Weekend Warriors   197
Carl M. Maresh, Jaci L. VanHeest

Progressive Heat Acclimatization 199
The Hours Before Exercise in the Heat 201
During Exercise 202
The Hours After Exercise in the Heat 204
Summary 205

Appendix A: Position Stands From the ACSM and the NATA   207
Appendix B: The Body's Responses to Various Heat Illnesses   215
Glossary   220
References   225
Index   263
About the Editor   271
About the Contributors   273

FOREWORD

Following the widely publicized heatstroke death of a professional football player, national attention has focused on the exertional heat illnesses among athletes. This attention is warranted because exertional heatstroke and exertional hyponatremia, most often experienced during summer training or competitive events, are medical emergencies. Physicians, nurses, emergency medical technicians, coaches, athletic trainers, athletes, industrial supervisors, military leaders, and soldiers can benefit from learning about the heat illnesses. Thus, *Exertional Heat Illnesses* is both necessary and valuable.

Although several clinical and scientific articles have reviewed the exertional heat illnesses, no book has focused solely on these disorders. Yet physicians, coaches, and event directors involved with exercise or labor in a hot environment will eventually encounter these exercise-associated disorders. This volume provides necessary information for these individuals and will spur exercise physiologists to study and clarify the causes and mechanisms of the exertional heat illnesses.

The contributors to *Exertional Heat Illnesses* were selected by the editor because they collectively possess decades of clinical and physiological training; they have medically treated the heat illnesses; they have published original research studies on etiology, treatment, and recovery of heat illnesses; they have chaired relevant position stand committees of the American College of Sports Medicine and the National Athletic Trainers' Association; and they are recognized as authorities by professional organizations. Their expertise is displayed with clarity and style in every chapter. The tables and figures in each chapter convey conceptual summaries that will guide future professionals for years to come. This exceptional book is vital reading for all individuals who challenge their bodies in the heat and for all coaches and health professionals who care for them.

E. Randy Eichner, MD, FACSM

Professor of Medicine, University of Oklahoma Medical Center
Team Internist, University of Oklahoma Sooners football team and other varsity athletics

PREFACE

The great deserts of the earth present ominous obstacles to human passage and progress. Covering approximately one-fifth of the earth's surface, their area is immense. Twenty-one deserts, for example, exceed 20,000 square miles. Few humans inhabit or exercise in these regions because of high air temperatures and sparse water supplies. Animals and plants also find deserts inhospitable. Only those species that have developed specialized abilities to maintain thermal, water, and salt balance survive.

In contrast, jungles and rainforests are fountains of life. Located adjacent to rivers in tropical regions, water abounds and dense vines, shrubs, and small trees flourish. This lush climate encourages more animal and plant species than does any other environment on our planet. But the virtually impenetrable terrain is less welcoming to humans, and athletes and laborers experience increased cardiovascular strain, difficulty in maintaining normal body temperature, and a lower capacity for work and exercise in such formidable tropical climates.

But these environments are the extremes. Most humans never live, exercise, or work in arid deserts or steamy jungles. Only 30 to 35 percent of the world's population lives between the tropic of Cancer and the tropic of Capricorn. In the United States, exercise in hot environments typically is seasonal. Maximum daily temperatures in 1999 exceeded 32°C (90°F) in four major northern cities—Philadelphia, Chicago, Indianapolis, and Rapid City—on only 18 to 31 days, respectively. Even in the four southern cities of Miami, New Orleans, Little Rock, and Dallas, peak daily temperatures exceeded 32°C (90°F) on only 58 to 100 days, respectively. However, just because hot weather is seasonal does not mean it should be dismissed.

Many Americans acknowledge that exercise is good for health, weight loss, and stress reduction. However, they often do not fully appreciate the stress that a hot environment places on their bodies, and they know little about the increased risk of illness or injury when exercising in hot environments. Because the popularity of exercise and sport continues to grow, the number of individuals at risk for heat illnesses increases each year.

The human body responds to stress by mobilizing nutrients and initiating protective responses in various organs. These responses successfully maintain equilibria (e.g., of temperature, fluid–electrolyte levels, blood pressure, pH, and blood constituents) in cool environments. But heat stress combined with exercise stress may overwhelm the body's organ systems. The purpose of this book is to describe the various ways that the human body is overwhelmed during exercise–heat exposure. In medical terms, these conditions are known as the *exertional heat illnesses.*

The following chapters describe ways to identify, treat, and prevent the heat illnesses that are most commonly experienced by athletes, recreational enthusiasts, and laborers: heatstroke, heat exhaustion, heat syncope (e.g., fainting), heat cramps, and hyponatremia. Most clinicians and athletes know something about the exertional heat illnesses, but few have a complete knowledge base to help them make informed decisions regarding exertion in hot environments or treatment, should a heat illness occur. Exertional heat illnesses occur in unique settings and individuals. This book is the first to focus exclusively on this topic.

The authors of this book have decades of laboratory and medical experience. Their descriptions represent a distillation of more than 70 years of clinical and scientific studies that they and other authorities have published.

Exertional Heat Illnesses provides advice to individuals in a variety of medical, academic, and commercial settings. Athletic trainers, physicians, nurses, and emergency medical technicians will find effective treatment options for all of the exertional heat illnesses. Instructors and students who are interested in environmental exercise physiology will find this a valuable textbook for special topics and seminar courses that require key references for advanced study. Coaches, athletes, industrial supervisors, and military leaders will learn the causes of heat-related illnesses and ways to prevent them. Fitness, conditioning, and training specialists who direct the physical activity of others will gain useful information for their clients and students.

Each exertional heat illness is examined separately. Chapter 1 explains the basic physiological principles underlying the body's control of temperature, water, and ions. Chapter 2 provides an overview of the exertional heat illnesses, as presented in the International Classification of Diseases. Chapters 3 to 6 each describe the physiological and clinical nature of a particular exertional heat illness; the signs, symptoms, and keys to recognizing the illness; and the first aid and treatment for the illness. Chapter 7 explains the minor heat illnesses that usually are not related to exercise.

Chapter 8 explores the physiological, psychological, environmental, and host factors (e.g., concurrent illness, medications, and uniforms) that predispose humans to the exertional heat illnesses. Chapter 9 summarizes the medical supplies, treatment protocols, staffing issues, and facilities involved in providing medical care for athletic events. Chapter 10 outlines practical approaches that athletes can use to optimize performance and health during recreational endeavors, training, and competition in hot environments. Adjustments that can be made in training patterns, eating habits, fluids, clothing, sleep, and medications are described. Appendix A summarizes the current accepted standards and recommendations of two prominent health and sports medicine organizations: the American College of Sports Medicine and the National Athletic Trainers' Association. These position statements provide a wealth of information regarding the exertional heat illnesses and fluid replacement. Appendix B summarizes the body systems that are affected by the exertional heat illnesses.

The purposes of this book will be fulfilled if readers maintain optimal health and performance when exercising in hot environments and ensure that these illnesses are treated promptly and properly when they inadvertently occur.

Chapter

Physiology of Heat Stress

John W. Castellani, PhD

When a person begins to exercise in a hot environment, myriad physiological responses occur to maintain exercise performance. Almost all of the body's organ systems are involved, including but not limited to the central nervous and cardiovascular systems. These physiological responses are under nervous and hormonal control. The combination of exercise and high environmental temperatures creates immense stress on human physiological responses. In fact, Rowell (1) stated that "probably the greatest stress ever imposed on the human cardiovascular system (except for severe hemorrhage) is the combination of exercise and hyperthermia." The combination of exercise and high environmental temperatures is not only a significant stressor on physiological control mechanisms, but it can also be life threatening.

This chapter provides a general overview of the physiological responses to exercise–heat stress. An appreciation of the normal response to exercise–heat stress and potential modifiers of this response will help readers understand the pathophysiological responses presented in subsequent chapters.

HEAT BALANCE

Changes in body core temperature are simply caused by either a positive or negative change in heat storage. If the body produces more heat than what it dissipates, body temperature rises. Conversely, if the body produces less heat than what is lost to the external environment, heat storage will be negative and core temperature will eventually fall. We can present these relationships between production and loss mathematically as follows (2):

$$\dot{S} = \dot{M} - (\pm\,\text{Work}) \pm \dot{E} \pm \dot{R} \pm \dot{C} \pm \dot{K}$$

where S = heat storage, M = metabolic heat production, E = evaporation, R = radiation, C = convection, and K = conduction. Positive numbers indicate heat gain, and negative values indicate heat loss. E, R, C, and K are the heat exchange pathways (3).

1

Physical exercise can increase whole-body metabolism by as much as 15 to 20 times the resting rate in healthy young males. But because exercise only uses 20% of this increased metabolism to produce useful work (i.e., 20% efficiency), the balance of the increased metabolism is given off as heat. This heat must be dissipated to the environment. If it is not, the core temperature will rise to high levels very early in the exercise bout. Exercise metabolism contributes only to heat gain.

Heat loss occurs through radiation, convection, conduction, and evaporation. **Radiation** is the movement of heat via electromagnetic waves. In the case of heat loss from the body, heat moves down its thermal gradient from high to low. For example, heat is lost from the body to the environment when the skin temperature exceeds air temperature. **Convection** is the movement of heat down its thermal gradient from an object to a surrounding liquid medium. In the case of an exercising person, heat may move from the body to the "liquids" of air, water, or internal body fluids. Heat loss is about 2 to 4 times greater in water than in air of the same temperature. **Conduction** is the movement of heat between two objects that are in direct contact with each other. During exercise–heat stress, conduction—for example, from the feet to the ground surface—is usually negligible (less than 1% of total heat exchange). **Evaporation** is the loss of heat through the process of changing sweat from a liquid phase to a gaseous phase. This phase change requires the input of heat from the skin, that is, the latent heat of vaporization of water. Sweat that merely drips off the body provides no cooling potential, because no heat was utilized to evaporate the sweat.

The mechanism of heat loss is dependent on the ambient temperature. At low ambient temperatures (5-10°C; 41-50°F), dry heat loss (i.e., heat loss from radiation and convection) is greater than heat loss through evaporation. At high ambient temperatures, evaporative heat loss predominates. High ambient water vapor pressure (high relative humidity) also affects heat loss. The amount of sweat that moves from the skin to the environment is less when the air is laden with water vapor than when the environment is dry.

TEMPERATURE REGULATION AND THE BRAIN

Temperature regulation, like other physiological responses, involves complex coordination, primarily in the preoptic area of the anterior hypothalamus (PO/AH). A number of neural pathways extend from the hypothalamus to the brainstem, spinal cord, and sympathetic ganglia (4). Most physiologists believe that the human body regulates its temperature around a **set-point temperature.** If the PO/AH temperature increases above the set-point temperature (e.g., 37°C), heat loss responses are initiated. Conversely, if the temperature falls below this set-point, various heat gain responses occur.

Three types of neurons in the PO/AH coordinate various thermoregulatory effector responses (e.g., sweating, vasodilation) (4). Warm-sensitive neurons are activated typically at body temperatures above 37°C (98.6°F) and initiate responses that aid in heat loss (figure 1.1). Cold-sensitive neurons increase their firing rate when the PO/AH temperature falls below the set-point temperature. Temperature-insensitive neurons do not increase their firing rate as a function of temperature fluctuations, but rather they modulate the responses of both warm- and cold-sensitive neurons. PO/AH neurons (primarily warm- and cold-sensitive neurons) integrate not only central but also peripheral information by projecting their dendrites to areas that are innervated by afferent neurons. This explains how input from the periphery may affect temperature regulation. During exercise–heat stress, about 90% of thermoregulatory effector responses

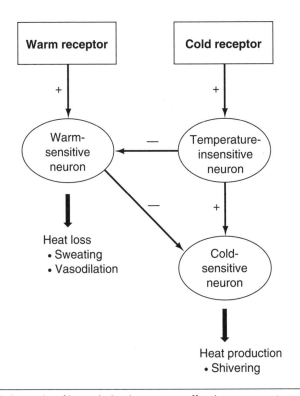

Figure 1.1 Schematic of hypothalamic neurons affecting temperature regulation.
Adapted, by permission, from J.R. Boulant, 1997, *Fever: Basic mechanisms and management*, edited by P.A. Mackowicik (Philadelphia: Lippincott-Raven), 35-58.

are responses to central changes (core temperature) and about 10% are responses to peripheral (skin temperature) changes (5).

MODIFIERS OF CORE TEMPERATURE REGULATION

Human beings originated in tropical regions near the equator and have, over a period of tens of thousands of years, migrated toward the polar regions. From a physiological perspective, humans adapt fairly well to hot climates but only moderately so to cold climates (6). The human body regulates its temperature between about 35 and 39°C (95-102.2°F). Normal core body temperatures fluctuate within this range depending on a number of factors, including circadian rhythms, menstrual cycle, endocrine disorders, and infections.

Circadian Rhythms

Core temperature fluctuates throughout a 24-hour period. The highest body temperatures are found at the end of the normal active period of each day, at about 1700 to 1900 hours, or 5 to 7 P.M. On the other hand, the lowest core temperature occurs during sleep, around 0400 to 0600 hours, or 4 to 6 A.M. (7). The difference in core temperature between the highest and lowest point of the day is typically about 0.5 to 1.0°C (0.9-1.8°F). Even if people are confined without external cues (daylight) and are inactive, the circadian rhythm of core temperature persists (7).

Menstrual Cycle

Basal body temperature is different during various phases of the menstrual cycle (8). During a normal 28-day menstrual cycle, body temperature is lowest during menses and the follicular phase (days 1 to 10). During this phase, from around day 5 to day 13, blood estrogen levels rise. Progesterone levels, on the other hand, do not change during this time period. Around day 13 or 14, just before ovulation, the pituitary secretion known as luteinizing hormone (LH) increases greatly. Following the "LH surge" and the subsequent ovulation, estrogen levels fall, but progesterone concentrations increase to a peak; this peak occurs around day 20 or 21. The phase between ovulation and the onset of menses is called the luteal phase. Basal core temperature changes concomitantly with these hormone fluctuations. During the follicular phase, when estrogen levels are higher relative to progesterone, core temperatures are low. However, during the luteal phase, when progesterone levels are high relative to estrogen, core temperature increases by about 0.5°C (0.9°F) and peaks between day 19 and 24. This change (increase) in set-point temperature has been used for many years as a means of natural birth control because it is tightly linked to ovulation and the transition to the luteal phase.

Endocrine Disease and Infection

Endocrine disorders and disease also affect core body temperature. For example, myxedema is characterized by a large increase in thyroid hormones and an elevated basal core temperature. Conversely, thyrotoxicosis is characterized by low thyroid hormones and lower than normal core temperature (9). Infections, whether bacterial or viral in nature, also increase core temperatures. This increase is commonly called a fever. The mechanisms underlying elevated temperatures caused by fever are different than those caused by exercise.

Fever is an elevation of the set-point temperature, caused by an infectious disease. During a fever, the mechanisms of heat loss (radiation, convection, and evaporation) and heat gain (metabolism) are still present, but they operate at a higher internal set-point temperature. Thus, sweating and cutaneous vasodilation are still initiated during exercise, but the temperatures at which they begin are proportional to the increased set-point. For example, if a person with a fever exercises, he or she will start out with a higher-than-normal core temperature, begin to sweat and experience dilation of the blood vessels in the skin at a higher core temperature, and have a higher-than-normal temperature when exercise ends. One might think of this as turning a thermostat from 20 to 23°C (68-73.4°F). The furnace still turns on the same way, but now it is regulating the temperature at a higher level.

Fever is caused by a virus or bacteria that stimulates the release of a host of chemical substances, known as pyrogens, which cause the PO/AH to reset to a higher temperature. When the brain senses that the person's temperature is lower than the new set-point temperature, it sends neural signals to increase heat production (mostly by increased shivering) and decrease heat loss. This causes the body temperature to rise until it matches the new set-point temperature. During the period in which the PO/AH set-point has changed but the core temperature has not yet reached this level, a person usually experiences some form of chills. Some authorities believe that elevated body temperature during a febrile episode enables the body's immune system to fight the infectious agent.

The increase in body temperature during exercise is not a change in the set-point temperature as occurs with fever. Instead, it is caused by an increase in metabolic heat production that is not dissipated, at least immediately, by radiation, convection, and evaporation. The increase in heat production is rapid, but heat loss mechanisms turn on at a slower rate (5). The difference between the two equals the increase in heat storage and core temperature. When heat production greatly exceeds heat dissipation, hyperthermia and the clinical emergency heatstroke may occur.

CONTROL OF WATER AND IONS

Total body water (about 42 liters) makes up approximately 50 to 70% of the total weight in a normal human (10). Many of the body's physiological and biochemical processes take place within this watery environment. Measuring changes in body weight is the easiest way to estimate changes in total body water. Water balance is achieved through a number of integrated regulatory processes. Water intake from drinking fluids and eating food is balanced by water output primarily through urine production, respiratory loss, and sweating.

Sixty-seven percent (28 liters) of the total body water is located in the intracellular space, whereas 33% (14 liters) is located in the extracellular compartment (11). The extracellular compartment is further divided into the **interstitial space** (10.5 liters) and the **plasma volume** (3.5 liters). Fluid is freely exchanged between these three compartments, depending on the ion and protein concentrations of each space. Fluid moves between the plasma and interstitial space via changes in **Starling forces** (10). On the other hand, fluid moves between the interstitial and intracellular spaces via **osmotic forces** (11). There are four Starling forces that either increase or oppose filtration of fluid out of a capillary into the interstitial space. The two forces that increase filtration out of a capillary are **capillary hydrostatic pressure** and **interstitial oncotic pressure** (caused by increased protein concentration). Higher pressures within the capillary "push" water out of the capillary and into the interstitium. Likewise, higher protein concentrations in the interstitial space draw fluid out of the plasma.

The two forces that oppose fluid moving into the interstitium are **plasma oncotic pressure** and **interstitial hydrostatic pressure.** Lower pressures within the capillary (relative to the interstitial space and primarily toward the venule end) cause fluid to move into the plasma (figure 1.2). Likewise, if protein concentrations are higher in the plasma, relative to the interstitial space, fluid will go into the plasma.

Any change in these four forces will lead to fluid exchange between fluid compartments. Body water is in a dynamic state; it is constantly moving between each of these spaces, depending on the forces acting on local capillary beds.

Sodium

Sodium regulation is closely tied to water balance. The adage "where sodium goes, water will follow" explains the general principle. Water moves from areas that have a low sodium or ion concentration to compartments that have a higher **osmolality.** Sodium is the most

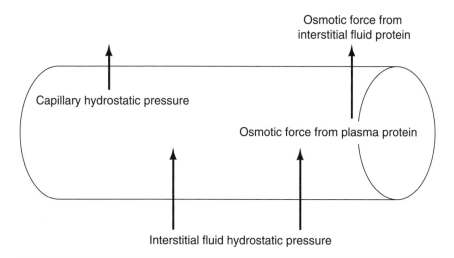

Figure 1.2 Different factors (or Starling forces) affecting fluid movement into and out of a capillary.

abundant ion in the extracellular fluid. It remains primarily outside cells caused by the action of the sodium-potassium pump, which moves sodium out of cells and potassium into cells.

Sodium intake in a typical American diet is around 10.5 grams per day. Because we essentially excrete the same amount in the urine, we maintain sodium and water balance by changing urinary output. If water and salt intake are low, water and salt excretion are low; likewise, a large consumption of water and salt proportionally increase urinary water and sodium losses.

Hormones

The hormone arginine vasopressin (AVP), also known as antidiuretic hormone (ADH), is the major determinant of water excretion (8). AVP is secreted from the pituitary gland and acts on the distal tubule and collecting ducts in the kidney, where it causes water to be reabsorbed. Low-pressure cardiopulmonary receptors and changes in the plasma osmolality regulate AVP. Dehydration is a potent stimulator of plasma vasopressin release.

Aldosterone, a steroid hormone produced in the adrenal gland, is the hormone that primarily regulates sodium reabsorption (8). When aldosterone is not present, large amounts of sodium are lost in the urine, whereas sodium losses are negligible in the presence of this hormone. Changes in the enzyme renin regulate aldosterone. Renin is secreted from the kidney and is primarily controlled through changes in kidney sympathetic nerve activity and pressure receptors within the kidney. Typically, low plasma sodium levels are associated with low plasma volumes. A low plasma volume, in turn, causes a reflex response that activates baroreceptors and increases sympathetic nervous activity that causes the kidney to secrete renin. Increased renin activity causes the formation of angiotensin II, a potent vasoconstrictor. Angiotensin II then acts on the adrenal cortex to increase the formation and secretion of aldosterone. The increased aldosterone causes more sodium to be reabsorbed into the bloodstream. As it does with vasopressin, dehydration stimulates renin, angiotensin, and aldosterone secretion to conserve total body water.

Exercise and Fluid Changes

In exercise physiology studies, plasma volume (PV) is commonly measured as a percent change from a baseline measure. Therefore, the equation developed by Dill and Costill (12) utilizing hemoglobin concentration and hematocrit is most often used. Some factors that may confound interpretation of the percent change in PV include posture, arm position, temperature, exercise mode, exercise intensity, exercise duration, hydration status, heat-acclimatization level, and fitness level (13). For example, because 70% of blood is below the heart level, standing up can reduce plasma volume by 10 to 15%. Preexercise skin temperatures also affect subsequent plasma volume measurements during exercise–heat stress. If the skin is cool during the preexercise blood sample, measured plasma volume will be lower and subsequent exercise–heat stress is likely to lead to an interpretation of increased plasma volume (13).

Cycling exercise causes an initial loss of PV that is mediated by hydrostatic forces (11). This loss of PV occurs without fluid loss (14). As the exercise duration in a hot environment increases, PV decreases as fluids are lost in sweat. This loss of PV (**hemoconcentration**) is associated with a loss of total circulating protein.

During treadmill walking, **euhydrated** (normally hydrated), heat-acclimated individuals typically hemodilute (gain PV) during exercise–heat stress at low exercise intensities. At higher exercise intensities, PV may not change for these individuals. However, when **hypohydration** is combined with exercise–heat stress, subjects experience hemoconcentration and lose protein from the circulation (15). During such conditions, PV starts to decrease at 3% and 5% losses of body weight both before and during treadmill exercise. This occurs even though studies suggest that preexercise total circulating protein increases with hypohydration.

PHYSIOLOGICAL RESPONSES TO EXERCISE–HEAT STRESS

Exercising in the heat causes profound thermoregulatory, cardiovascular, and metabolic changes. Without these acute physiological responses, the ability to work would be curtailed and the likelihood of heat injuries would increase.

Thermoregulatory Responses

During exercise, the increase in core temperature during compensable heat stress (i.e., where the maximal evaporative capacity in the environment is greater than the evaporation required by the individual to maintain body temperature) is independent of the environmental conditions (16) but is related to the metabolic rate. This relationship is true in a range of environmental temperatures known as the "prescriptive zone." Between about 10 and 25°C (50-77°F), steady state core temperature is the same when exercising at a fixed metabolic rate, no matter what the air temperature is. The core temperature rises within this zone only when exercise intensity increases, although the range of ambient temperatures within the prescriptive zone decreases as the exercise intensity increases. However, above ambient temperatures of 25°C (77°F), steady state core temperature rises proportional to the increase in ambient temperature.

The increase in core temperature during exercise is dependent on the absolute metabolic rate within an individual. However, between persons, exercise at a given absolute metabolic rate elicits large interindividual variability. Astrand (17) found that between individuals, core temperature responses are similar when expressed as a percentage of the maximal oxygen uptake ($\dot{V}O_2$max). This concept is important because, when comparing two people of varying fitness levels who are exercising at the same absolute workload, the person who is less fit will have a higher steady state core temperature. Thus, in a group of athletes or soldiers who are running at the same pace, the individuals who have lower maximal aerobic power will experience greater heat storage.

Cardiovascular Responses

To effectively dissipate heat during exercise, blood must be shunted from the core to the periphery (skin) for heat loss via radiation, convection, and evaporation. How does this occur, and what are the consequences of this on the rest of the cardiovascular system? Exercise–heat stress increases venous compliance, particularly in the skin, which causes a redistribution of blood to the periphery and lowers central blood volume and, hence, venous return to the heart. Lower venous return decreases cardiac filling and stroke volume. To maintain an adequate cardiac output during exercise, heart rate subsequently increases to offset the decreased stroke volume (1). These changes are magnified during upright exercise, when even more blood pools in the lower extremities because of gravity. Eventually, if not enough blood is returned to the central circulation, **syncope** may result.

Exercise independently causes constriction of the blood vessels in the **splanchnic** (gut, liver, and kidney) region and in the cutaneous circulation (1). Capillaries in active skeletal muscle dilate to meet the metabolic demand. Heating also independently leads to splanchnic vasoconstriction. However, in response to heating, skeletal muscle blood vessels constrict and the cutaneous circulation vasodilates to increase blood flow to the periphery (1). This redistribution of cardiac output in response to heating results in a drop of central venous pressure. When exercise is combined with heating, competition for blood flow occurs between active skeletal muscle and skin. Skin blood flow is sacrificed to maintain muscle blood flow, and a higher core temperature results. When cardiac output cannot meet the demand of muscle metabolism, muscle blood flow normally will be sacrificed to maintain arterial pressure (1, 18).

Metabolism and Substrate Use

Metabolism during exercise–heat stress, compared to exercise in a temperate environment, may increase anaerobic metabolism (19). For example, Dimri et al. (20) found that the percentage of energy derived from aerobic and anaerobic metabolism decreased and increased, respectively, as the ambient temperature increased from 27 to 40°C (80.6-104°F). This increased reliance on anaerobic metabolism is reflected in a higher blood lactate concentration. Lactate levels in muscle also have been reported to be higher at high ambient temperatures than at low ambient temperatures. The mechanism for the higher lactate levels may be increased muscle glycolysis.

Fink, Costill, and Van Handel (21) found that muscle glycogen was depleted to a greater extent during exercise at 41°C (105.8°F), compared to a 9°C (48.2°F) environment. Higher respiratory exchange ratios ($\dot{V}CO_2 / \dot{V}O_2$, an index of type of substrate utilized) also have been documented, suggesting a shift to carbohydrate use in hot (versus cool or temperate) environments. It is likely that the shift in substrate utilization occurs when the core temperature rises more than 0.5°C (0.9°F). However, during exercise to exhaustion in the heat, muscle glycogen levels are not depleted, as they are when exercising at cooler temperatures (22), even though glycogen utilization is greater during exercise–heat stress. These data suggest that factors not related to substrate utilization are responsible for fatigue during exercise–heat stress. Several potential hypotheses need to be tested, including decreased central nervous system drive and increased intramuscular inosine monophosphate (IMP) levels, which reflect a temperature-induced effect on metabolism (22). Interestingly, heat acclimation has been shown to lower lactate levels and reduce the body's reliance on muscle glycogen as a substrate.

FACTORS THAT LIMIT EXERCISE PERFORMANCE IN HOT ENVIRONMENTS

Several mechanisms have been proposed to account for reduced exercise performance during prolonged exercise–heat stress. Exercise in the heat can cause substantial body water loss and a resultant dehydration. This reduces stroke volume. If heart rate does not increase to offset the lower stroke volume, then cardiac output also will decrease, limiting oxygen and substrate delivery to the working muscle. Therefore, the heart rate increases, causing cardiovascular strain. This chain of events may eventually lead to heat exhaustion, which will be discussed fully in chapter 4. Exercise–heat stress, even without dehydration, causes cardiovascular drift. Over time, heart rate increases steadily until the high heart rates, in effect, lead to exhaustion.

More recently, it has been proposed that fatigue during exercise–heat stress occurs when a "critical core temperature" is achieved (23). Evidence for this comes from studies by Neilsen and colleagues (24, 25) in Denmark. They found that exhaustion during exercise–heat stress occurred at the same body core temperature during a heat-acclimation regimen. Similarly, exhaustion occurred at the same body core temperature during exercise that began at various core temperatures. For example, exercise endurance improved 67% during heat acclimation, but the subjects fatigued at the same core temperature each day. However, other studies (26, 27) do not support this finding; instead, they found that core temperatures at exhaustion range from 38 to 40°C (100.4-104°F) depending on if the subjects were hydrated, wearing clothes, or exercising at different intensities. It may be, in fact, that cardiovascular strain and the attainment of one's maximal heart rate determines the point of exhaustion. For example, in one of the Danish studies (25) investigators observed that, in addition to body core temperature, heart rates and cardiac output were similar at the point of exhaustion across different trials (heat acclimated or not, starting at various body temperatures). Thus, there is still some disagreement on the mechanism that limits performance during exercise–heat stress. It also is likely that performance is limited by different factors when exercise involves different modes, intensities, or durations.

FACTORS THAT MODIFY
THE EXERCISE–HEAT STRESS RESPONSE

Several factors may modify, either positively or negatively, the thermoregulatory and cardiovascular responses to exercise–heat stress. These include heat acclimation, hypohydration, gender, age, and spinal cord injury.

Heat Acclimation

Repeatedly exposing a person to a hot environment lowers thermoregulatory and cardiovascular strain and improves exercise performance (28). Becoming heat acclimated requires 10 to 14 days with one exercise session per day of continuous 100-minute heat exposure, although there is evidence that shorter, more intense exercise also induces heat acclimation. A prolonged increase in core and skin temperature is the most important factor for inducing heat acclimation. The time course of acclimation varies depending on the organ system. For example, heart rate is lowered within 2 to 4 days. Improved sweating responses are typically observed after only 2 days, but require 10 to 14 days to reach their peak. A reduction in core and skin temperature also occurs after 3 to 4 days.

Heat acclimation is induced most quickly during exercise–heat stress, but it also may be induced partially via resting heat exposures. Being physically fit also improves exercise–heat tolerance, but not fully. Even persons who are extremely fit need to exercise in the heat to become fully acclimated (29).

Evidence of heat acclimation includes lower heart rates, lower core and skin temperatures, and improved sweating rates during exercise and reduced sodium losses in sweat and urine (30). Sweating also begins at a lower core temperature threshold when a person is heat acclimated. Similarly, skin blood flow is higher at any given core temperature in a heat-acclimated person, because of a change in the threshold temperature at which cutaneous vasodilation begins to rise. Thus, the lower skin temperatures following heat acclimation reduce the volume of skin blood flow for heat exchange, but acclimation initiates skin blood flow sooner so that dry heat loss (radiation and convection) improves to an even greater extent. Higher sweating rates and earlier skin blood flow, in concert, help explain why core temperature is lower in a heat-acclimated person; heat-acclimated people have better evaporative, radiative, and convective heat loss than those who are not acclimated.

An improved sweating response is the most critical factor for heat acclimation and health (11). A better sweating response improves evaporative heat loss and lowers skin temperature. A lower skin temperature decreases skin blood vessel compliance (i.e., increases peripheral resistance), which in turn helps redistribute blood back to the central circulation and lowers cardiovascular strain. Because heat-acclimated individuals have high sweating rates, their fluid requirements are greater and they are more susceptible to dehydration than their non-heat-acclimated counterparts.

Acclimating to humid heat (high relative humidity) may be somewhat different than acclimating to dry heat (low relative humidity). Acclimating to humid heat may require a higher skin blood flow (improving radiative and convective heat loss) and sweat evaporation on a greater area of the body than observed during dry heat acclimation. Although being acclimated to hot–dry conditions improves exercise–heat tolerance during exercise in humid heat, it does not help a person fully tolerate such conditions (6). Thus, it is prudent for athletes to practice in the environment that they likely will encounter during competition.

Hypohydration

Hypohydration and dehydration are deleterious to exercise performance and health in hot environments. Cardiovascular and thermal strain are greater when hypohydrated than when euhydrated. For example, when exercising when hypohydrated, compared with exercising when euhydrated, sweating doesn't begin until a higher core temperature is reached, skin blood flow is lower, and heart rates are higher. Also, the thermal advantages (i.e., lower core temperature) gained by heat acclimation and aerobic training are lost (6, 11) when exercising in a hypohydrated state. Therefore, it is important to consume fluids before and during exercise in the heat (31).

Gender

Studies have shown that exercise–heat stress responses are not different between men and women, as long as they have similar aerobic fitness levels and body composition. However, because women generally have a lower $\dot{V}O_2$max than do men, exercising at the same absolute exercise intensity leads to a higher core temperature in women, because they are exercising at a higher relative exercise intensity. Also, because women have a smaller average volume of total body water, loss of a given volume of sweat (e.g., 3 liters in 2 hours) results in greater percentage fluid loss as compared to fluid loss in men.

Compared to men, women have a greater percentage of fluid loss. Drinking fluids during exercise can help alleviate fluid loss.

Age

Various findings have been reported regarding thermoregulation in children. In moderate to moderately hot ambient temperatures (20-35°C; 68-95°F), the thermoregulatory responses of adults and children are not different. However, in very hot environments (45°C; 113°F) children have a lower heat tolerance.

Children, compared to young adults, have a higher surface-area-to-mass ratio. In warm environments, this is an advantage because it promotes radiative and convective heat loss. But when the ambient air temperature is much greater than skin temperature, children absorb more heat via dry heat exchange because of their higher surface-area-to-mass ratio.

It is typically assumed that thermoregulatory responses to heat are reduced as one ages. However, a careful analysis of the research literature indicates that many of the differences between younger and older adults are due to differences in physical fitness. In the general population, exercise is more stressful in the heat in older individuals. This is because the maximal aerobic fitness, in general, declines as one ages. However, in the studies that have matched older and younger adults for $\dot{V}O_2$max, the cardiovascular and thermoregulatory strain was not different between young and old (32). Both children and older adults can acclimate to the heat. For young children, the time course of acclimation may be a bit longer than for young adults. Acclimating to heat and improving exercise performance and thermal comfort is common throughout each stage of life.

Spinal Cord Injury

The number of people with spinal cord injury (SCI) who exercise is increasing. These individuals' injuries certainly affect exercise performance during exercise–heat stress. Thermoregulatory impairment in people with SCI is related to the location and severity of the spinal injury (6). Essentially, thermoregulatory impairments are found in areas of the body that are innervated from the spinal cord below the level of the spinal lesion. In these regions, sweating rate and peripheral blood flow are lower. These changes decrease evaporative, radiative, and convective heat loss; increase thermoregulatory strain; and reduce exercise performance in hot environments.

SUMMARY

Human body temperature is regulated between 35 and 39°C (95-102.2°F) via various thermoregulatory effector responses, such as sweating, dilation and constriction of skin blood vessels, and shivering, that are controlled by the anterior hypothalamus. Exercise–heat stress is a potent stimulus for various physiological changes in the cardiovascular, endocrine, and central nervous systems. Two of the most important modifiers of the exercise–heat stress response are acclimatization and dehydration. Heat acclimation improves a person's ability to perform in the heat, whereas dehydration worsens performance and can have health consequences.

DISCLAIMER

The views, opinions, and findings contained in this chapter are those of the author and should not be construed as an official Department of the Army position, policy, or decision unless so designated by other documentation. Approved for public release; distribution is unlimited.

Chapter

Classification, Nomenclature, and Incidence of the Exertional Heat Illnesses

Lawrence E. Armstrong, PhD, FACSM

Clearly Defined Illness Categories Aid Diagnosis

Case Report 2.1

On a hot–dry (40°C, 104°F; 15% relative humidity) summer morning, V.R., a 35-year-old man, worked outside an open warehouse for 4 hours at a mild, intermittent pace. During his lunch period, he complained of exhaustion, but he subsequently worked for another hour. After leaving work because he wasn't feeling well, V.R. traveled home via public transportation, began shouting angrily at nearby commuters, and exhibited bizarre behavior. A group of concerned passengers subdued him and summoned police, and V.R. was transported to a local hospital. Commuters told police officers that he had "acted like a crazy man." In the emergency department, V.R.'s heart rate was 130 beats per minute, blood pressure was 110/70 mmHg, and ear canal temperature was 37.5°C (99.5°F). V.R. was treated as a potential psychiatric patient because of frequent outbursts, incoherent speech, and combative behavior. He was sedated and isolated for 2 hours of observation while a psychological evaluation was scheduled. V.R., usually healthy, had no history of mental illness.

Rectal temperature was not measured until V.R. vomited and lapsed into unconsciousness. The measurement of 42°C (107.6°F) alerted nurses that the patient was hyperthermic. External cooling was applied and his core body temperature returned to normal. The neurological examination found V.R. in a deep coma (Glasgow coma scale = 8) with eye deviation upward, diminished deep tendon reflexes, and normal tonus. A preliminary diagnosis of exertional heatstroke was made; the diagnosis was confirmed after the patient regained consciousness and recounted the events leading to his collapse.

An initial laboratory evaluation revealed that blood urea nitrogen (BUN) concentration was 32 mg · dL⁻¹ (normal range is 8-25 mg · dL⁻¹) and electrolyte concentrations were normal. After 24 hours of hospitalization, disseminated intravascular coagulation (DIC) with a platelet count of 19,000/μL, fibrinogen 212 mg · dL⁻¹, and prolonged prothrombin clotting time were observed. Serum creatine phosphokinase levels were mildly elevated (750 IU · L⁻¹), without myoglobinuria. After 3 days of hospitalization, V.R. was alert and well oriented and exhibited normal neurological function; all blood factors were normal.

This case report illustrates the potential for misdiagnosis when emergency response teams and clinicians do not consider exertional heat illness as a possible diagnosis. In cases where patients have exhibited bizarre or combative behavior and the circumstances surrounding collapse, or personal history, have been unknown, exertional heatstroke has been misdiagnosed as psychotic behavior. Unambiguous descriptions for each of the exertional heat illnesses can prevent such misdiagnoses.

Because of requirements for heavy physical exercise in harsh environmental conditions, heat illnesses have been reported in military contexts for more than 2,000 years. In the year 24 B.C., a Roman officer named Aelius Gallus conducted a military campaign in Arabia. Not specifically trained for desert survival, the majority of his troops succumbed to intense heat, solar radiation, dehydration, and eventually heatstroke (1). In the 13th century, King Edward I and his heavily armored crusaders allegedly lost their final battle of the Holy Land, against well-acclimated Arab horsemen, because of heat and fever (2). During World War I, the British army lost many soldiers as heat casualties during a battle against Turkish forces in Mesopotamia (3). A summary of heat casualties among Indian military personnel during 1963 and 1964 reported that heat exhaustion, heat syncope, heat cramps, and heatstroke occurred primarily during the months of June and early July (4). Similarly, the incidence of U.S. heat casualties during World War II, the Vietnam conflict, and in basic training units during peaceful years indicates that heat illnesses were a significant source of morbidity and mortality during the 20th century (5).

Numerous cases of exertional heat illness and heatstroke have occurred in mass participation foot races (6) and cycling events in Australia (7), Canada (8), France (9), New Zealand (10), Norway (11), and the United States (12, 13). In addition, deaths caused by heat-

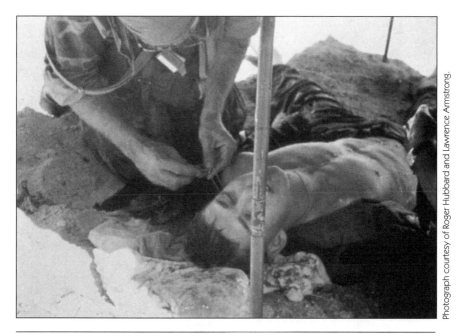

Photograph courtesy of Roger Hubbard and Lawrence Armstrong.

Exertional heat illnesses can be problematic for soldiers who are unaccustomed to working in the heat.

stroke historically have been of concern among American football players (14, 15).

In industrial settings, strenuous exercise in hot environments and large sweat losses have been notorious as causes of exertional heat illness aboard oil tankers in the Persian Gulf (16), in the mines of South Africa (17), and at the construction site for the Hoover Dam in Nevada (18).

Because heat disorders have been experienced in such diverse settings, reports of the symptoms of each illness may be subtly different. This makes diagnosis more difficult and stimulates debate regarding the etiology involved. The purpose of this chapter is to clarify the classification, nomenclature, and the incidence of the illnesses described in this book. These details will enhance the reader's ability to understand, identify, prevent, and treat the exertional heat illnesses.

CLASSIFICATION

Uniformity in reporting heat disorders is most useful when based on a widely accepted, current classification and nomenclature. The International Classification of Diseases (ICD), published by the World

Health Organization, offers one means to accomplish this objective. The disorders that involve environmental heat were first codified in 1957 by the Medical Research Council's Climatic Physiology Committee of the United Kingdom and the U.S. National Research Council's Subcommittee on Thermal Factors in the Environment (19).

Although the ICD originally included terms (e.g., sunstroke) that now are recognized as obsolete (20), subsequent revisions have streamlined this taxonomy. Presently, the ICD contains 10 categories of heat disorders (21), as shown in table 2.1, the first 10 listings. In addition, hyponatremia and diseases of the sweat glands are included here because they frequently occur in endurance events or during prolonged labor in hot environments. The heat disorder categories do not include sunburn, malignant hyperthermia following anesthesia, or other diseases of the sweat glands, which are classified under other disease categories. However, these disorders are included in table 2.1 because a hot environment plays a role in their etiology.

Table 2.1

Disorders in Which Hot Environments Are an Etiologic Factor

ICD category	Etiology
Heatstroke	Overwhelmed or failed thermoregulation
Heat syncope	Circulatory instability
Heat cramps	Water and electrolyte imbalance
Anhidrotic heat exhaustion	Water and electrolyte imbalance
Salt deficiency heat exhaustion	Water and electrolyte imbalance
Water deficiency heat exhaustion	Dehydration; hypohydration
Unspecified heat exhaustion	Unspecified causes; exertion
Transient heat fatigue	Psychological causes
Heat edema	Water and electrolyte imbalance
Unspecified heat effects	Various causes
Hyponatremia	Sodium deficit
Water intoxication	Excessive water intake
Miliaria rubra	Sweat gland obstruction
Sunburn	Ultraviolet radiation burn of the skin
Tropical anhidrotic asthenia	Psychological causes

Definitions for disorders appear in the text. ICD = International Classification of Diseases. Adapted from Center for Health Statistics, 1998, *International Classification of Diseases*, 9th revision, Clinical Modification (ICD-9-CM) (Hyattsville, MD).

In recent years, researchers have rarely disputed the classification of exertional heat illnesses (the first 10 disorders in table 2.1), with the exception of exertional heat exhaustion. Because the symptoms present in various ways in different situations and individuals, and because it is diagnosed by exclusion (i.e., other illnesses are eliminated until there is no other reasonable option) (22), some clinicians believe that exercise-induced heat exhaustion is merely orthostatic hypotension (T.D. Noakes, personal communication, June 2002; W.O. Roberts, personal communication, June 2002) and treat it as an exercise-induced collapse. Heat exhaustion and collapse were classified together in 1957 but were separated in 1965 (20). Other authorities (4) combine heat syncope and exertional heat exhaustion, with separate designations for water depletion heat exhaustion and salt depletion heat exhaustion. However, because there are no clinical or controlled laboratory studies to verify these taxonomies, the ICD categories in table 2.1, which are recognized throughout the world (23-27), will be used in this text. Although other taxonomies (28) and definitions (29, 30) exist, a more elaborate classification would have little practical value in the field, where the clinician must rely on clinical observations and judgments without the benefit of laboratory tests.

NOMENCLATURE

The following paragraphs provide concise definitions for each exertional heat illness described in this book. These definitions have been distilled from the ICD (21) and other reputable sources (19, 20, 25-27, 31, 32).

anhidrotic heat exhaustion—Synonym: prostration. Initially described during World War II and rarely reported since. Extensive obstruction of sweat gland ducts leading to impaired evaporation, impaired cooling, and heat intolerance. Almost always preceded by *miliaria rubra* (see below). Involves weakness, irritability, loss of initiative, difficulty concentrating on tasks, and insomnia.

heat cramps—Painful spasms of skeletal muscles, usually after exercise. Precipitated by large sweat (i.e., water and sodium) losses accompanied by drinking **hypotonic** fluids or pure water. Resolved by sodium replacement.

heat edema—Swelling in the extremities. Ranges from tightness of footwear or watchband to incapacitating swelling of the ankles and lower legs. Fluid pools in the interstitial space. Resolves with heat acclimatization or relief from heat stress.

heatstroke—Synonym: hyperpyrexia. Homeostatic and thermoregulatory mechanisms are unable to dissipate the heat produced during strenuous exercise. This disorder is characterized by hyperthermia (>39-40°C; >102.2-104°F) that causes profound central nervous system disturbances such as confusion, delerium, or coma; hypotension; convulsions; vomiting; diarrhea; disseminated intravascular coagulation; and elevated serum enzymes.

heat syncope—A brief fainting episode in the absence of salt and water depletion, fluid loss, or hyperthermia, often subsequent to prolonged standing. Results from decreased central venous return caused by pooling of blood in the legs and skin. Tunnel vision, vertigo, nausea, and weakness may precede syncope.

salt deficiency heat exhaustion—Synonym: prostration. Inadequate replacement of the sodium lost during sweating, leading to decreased extracellular and blood volumes. Urine sodium ion (Na^+) concentration is very low. Symptoms may include collapse, pallor, sweating, weakness, vomiting, and perhaps skeletal muscle cramps.

transient heat fatigue—Synonym: acute heat neurasthenia. Involves psychological symptoms such as extreme tiredness, disinclination to work, irritability, and errors in skilled performance, in the absence of salt and water deficiency and anhidrosis. Attributed to short-term exposure to extreme heat and poor ventilation. Symptoms are rapidly relieved when the heat stress is removed.

unspecified heat effects—Signs and symptoms not otherwise specified and caused by hot environments.

unspecified heat exhaustion—Occurs in the absence of salt and water deficiency and anhidrosis without other symptoms.

water deficiency heat exhaustion—Synonym: prostration. Attributable to inadequate water replacement in a hot environment. Characterized in the earliest stage by thirst, vague discomfort, anorexia, impatience, weariness, sleepiness, and dizziness. Restlessness, rapid breathing, cyanosis, and delirium eventually develop if left untreated.

A few other illnesses are observed in association with hot environments but are not categorized in the ICD with the previously mentioned heat disorders. These include the following disorders.

hyponatremia—Sodium (Na^+) deficiency that involves a serum or plasma sodium concentration of <130 mEq $\cdot$ L^{-1} (some authorities recognize 135 mEq $\cdot$ L^{-1} as the definitive level). Hyponatremia results from replacement of sodium losses in sweat and urine with hypotonic fluid or water. Severe cases may involve coma, pulmonary or cerebral edema, and death. Hyponatremia usually occurs when individuals incorrectly perceive that they need to consume a larger quantity of water than is needed (e.g., before and during exercise or labor in hot environments).

miliaria rubra—Synonym: heat rash or prickly heat. A superficial, fine papular eruption and redness of the skin that follows excessive sweating and plugged sweat glands, particularly in hot–humid environments. Often becomes infected or inflamed. Relieved in days to weeks by removal of thermal stress and regular skin cleansing.

tropical anhidrotic asthena (unclassified)—Synonym: chronic heat neurasthemia. A chronic neurotic or depressionlike illness, in the absence of a clearly defined organic heat disorder, attributable to prolonged exposure to a hot environment (see chapter 7). Involves loss of initiative and performance, difficulty in concentrating on tasks, insomnia, loss of weight and appetite, irritability, and depression. Some authors believe that this disorder is identical to neurosis or depression in cooler climates.

water intoxication—Consumption of an excessive volume of water that results in symptoms such as headache, dizziness, vomiting, convulsions, coma, and death.

INCIDENCE

Incidence is the frequency of occurrence of an illness over a period of time and in relation to the population in which it occurs. The incidence of each exertional heat illness is influenced by etiological factors, number of participants, population involved, and the stresses imposed by the event. For example, table 2.2 illustrates that the incidence of heatstroke is difficult to predict because it ranges from 0.1/1,000 individuals/3 months to 26.2/1,000 individuals/day. Similarly, hyponatremia involves a wide range of incidence, depending on the event. But heat exhaustion exhibits a relatively narrow incidence (e.g., <0.1/1,000 individuals/day to 1.4/1,000 individuals/day) in the events reported.

The combined incidence of all exertional heat illnesses, as reported by five authors, is presented in table 2.3. This compilation illustrates the wide range of incidence statistics (0.7/1,000 individuals/day to 57/1,000 individuals/year) that occur in different settings. This clearly makes it difficult to project the number of casualties in one situation to another scenario.

Figure 2.1 suggests that prediction of the number of exertional heat illnesses for specific events or populations will be most effective if medical records from previous years are consulted. Such estimates of occurrences will likely provide closer estimates of exertional heat illness incidence for future events and can guide advanced organization of resources, supplies, equipment, transportation, communication, support staff, and medical emergency response teams. However, even though recommendations exist regarding the number of these items that should be on-site at mass participation events

Table 2.2

Incidence of Exertional Heat Illnesses and Hyponatremia

Heat illness	Number, event (reference)	Incidence of illness[a]	Deaths (%)[b]
Exertional heatstroke	• 7,000 Marine recruits in training, southern U.S. (36)	0.2/3 mo	0
	• 17,632 road runners at 10-km summer road race (37)	0.2/d	0
	• Two million religious pilgrims; desert trek in Saudi Arabia (38)	0.6/wk[c]	4.7-18.4
	• 2,897 runners in 42.1-km road race at 31°C (87.8°F) (13)	26.2/d[d]	0
	• Marine recruits in basic training from 1975-1989 (39)	0.1/3 mo[e]	0
	• 216,615 Marine recruits in basic training from 1982-1991, southern U.S. (40)	0.4/3 mo	1.3
	• 7,100 road runners at 11.5-km summer road race (41, 42)	1.4/d	0
Heat exhaustion	• 7,000 Marine recruits in summer basic training (36)	0.3/3 mo	0
	• 6,010 reserve soldiers, summer maneuvers (43)	18.3/2 wk[c]	0
	• Two million religious pilgrims; desert trek in Saudi Arabia (44)	0.3-0.4/wk	0
	• 144,950 runners in 14-km summer road race, 1978-1984 (30)	1.4/d	0
	• 63,732 runners at 14-km summer road race (45)	0.1/d	0
Heat syncope[f]	——	——	——
Heat cramps	• Field laborers in equatorial British Guiana (46)	<10/y	0
	• Industrial and shipboard laborers and military personnel in the tropics (26)	<10/y	0

Heat illness	Number, event (reference)	Incidence of illness[a]	Deaths (%)[b]
Hyponatremia[g]	• 315 runners in 90-km ultra-marathon at 30°C (86°F) (47)	79.4/d	0
	• 101 runners in 186-km ultramarathon at >30°C (>86°F) (47)	0/d	0
	• 373 competitors in 226-km New Zealand Ironman Triathlon at 21°C (69.8°F) (48, 49)	0.3-1.7/d	0
	• 5,028 marathon runners in a 42.2-km race at 5-10°C (41-50°F) (50)	4.2/d	0

[a]Per 1,000 participants. [b]Percentage of the patients in column 3 that died. [c]The minimum incidence; numerous cases were not surveyed. [d]Extreme hyperthermia; some patients were hospitalized. [e]The total number of recruits was estimated from data in reference 40. [f]No known incidence data available. [g]Serum sodium concentration <130 milliequivalents per liter (mEq · L^{-1}).

d = day; wk = week; mo = month; y = year.

Table 2.3
Combined Incidence of Exertional Heat Illnesses

Number, event (reference)	Combined incidence[a]
World War II soldiers during desert duty in the Persian Gulf (51)	57/y[b]
Soldiers during daily duty in the Vietnam War (52)	0.7-5.4/d
6,010 reserve soldiers during summer maneuvers (43)	23.5/2 wk[c]
526 Marines during summer maneuvers (5)	62.7/2 wk[c]
216,615 Marine recruits in basic training from 1982-1991 in the southern U.S. (40)	6.7/3 mo[d]

[a]Per 1,000 participants. [b]Includes only hospitalized casualties. [c]Includes heatstroke, heat exhaustion, heat syncope, and heat cramps. [d]Includes heatstroke, heat exhaustion, heat syncope, heat cramps, and rhabdomyolysis.

d = day; wk = week; mo = month; y = year.

(34), an unexpected increase in the temperature, humidity, or solar radiation can markedly increase the number of heat illness cases. Indeed, the number of exertional heatstroke occurrences often rises precipitously during a heat wave because participants have not acclimated adequately before the event and because the hot environment reduces dissipation of metabolic heat.

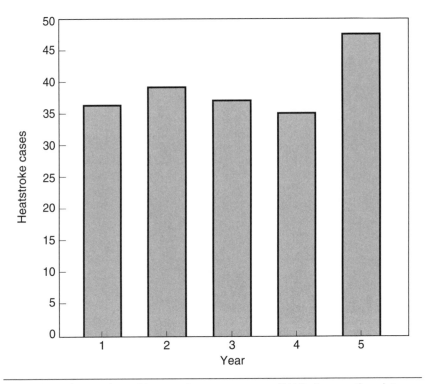

Figure 2.1 The number of heatstroke cases among all U.S. Army active duty personnel during a 5-year period.

The 1986 Pittsburgh Marathon provides a case in point (see data from reference 13 in table 2.2). More than half of the 2,897 competitors were treated for heat-related injuries, in 31°C (87.8°F) conditions. Along the 42.2-kilometer (26.2-mile) course, 450 runners received medical attention, primarily for hyperthermia. Fifty-one entrants were hospitalized for treatment of hyperthermia. In comparison, during the previous year's event, which had a maximum ambient temperature of 24°C (75.2°F), only 600 runners were treated for injuries of all types (e.g., exertional heat illnesses plus sprains, blisters, and orthopedic problems). Interestingly, 970 of the competitors in the 1986 marathon had not run a marathon before the 1986 event.

Inherent host factors (e.g., illness, poor fitness, sleep loss) also may result in a rapid, large increase in the incidence of heat exhaustion. One military training exercise in the Middle East illustrates this principle (35). In August 1983, members of an 82nd Airborne unit conducted cooperative maneuvers with Saudi Arabia in Berbera, Somalia. The air temperature routinely exceeded 33°C (91°F). During a morning training run of about 3.5 kilometers (2.2 miles), 24 cases of severe heat exhaustion occurred. This training run had been performed on previous mornings without incident. The difference on this particular morning was an outbreak of infectious dysentery, later identified as *Shigella* bacteria derived from impure water sources. The most common symptoms (with the percentage of soldiers displaying each) were nausea (90%), weakness (80%), diarrhea (80%), thirst (70%), headache (60%), and dizziness (50%). By afternoon (1700 hours) the number of exertional heat illnesses in this unit exceeded 50. This number of patients, appearing within a 9-hour span, severely taxed the battalion field aid station.

During the course of a month or year, the ratio of the number of exertional heat illness cases to total event participants remains relatively stable. Figure 2.2 illustrates percentages of exertional heat illnesses commonly observed during labor, outdoor competitive

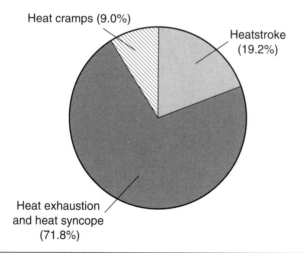

Heat cramps (9.0%)

Heatstroke (19.2%)

Heat exhaustion and heat syncope (71.8%)

Figure 2.2 Percentages of exertional heat illnesses reported during one year, among all active duty personnel stationed at military installations in the United States.

Reprinted from D. Arday, 1990, *Heat injury occurrence and prevention,* U.S. Army Research Institute of Environmental Medicine (Natick, MA).

events, and military maneuvers in hot environments. Typically, heat exhaustion is the most common heat illness, followed by either heatstroke or heat cramps. In fact, the latter two illnesses occur with similar frequency in both military and industrial settings (18). A discussion of heat exhaustion and heat syncope appears in chapter 4.

SUMMARY

The incidence of exertional heat illnesses is specific to the event, the severity of environmental heat stress, exercise intensity and duration, and inherent host factors. Because large differences have been reported in the incidence of some exertional heat illnesses (table 2.2) and because numerous factors contribute to hyperthermia (chapter 1), accuracy in predicting the incidence of exertional heat illnesses in a future event or comparison of the incidence of different exertional heat illnesses can best be accomplished by investigating the history of similar events. In addition, the incidence may increase unexpectedly if environmental heat stress increases or if detrimental host factors exist.

Chapter

Exertional Heatstroke: A Medical Emergency

Douglas J. Casa, PhD, ATC, FACSM
Lawrence E. Armstrong, PhD, FACSM

Exertional Heatstroke During a Mountain Bike Competition	Case Report 3.1

An enthusiastic 23-year-old male (K.G.) entered a cycling competition on a hot–humid day in August. Although he had little previous experience in mountain bike racing, he was well trained and maintained his fitness via a variety of exercise modes. According to three friends, K.G. ate a typical healthy meal the night before the race, showed no signs of having a cold or influenza, slept for 6 to 7 hours, and did not abuse alcohol or drugs.

Two hundred fifty racers entered this event, which began at 10:00 A.M. The course followed dirt paths that wound through a dense forest. The trails were not well marked, and about 4.8 kilometers (3 miles) into the race K.G. strayed off course; he was found unconscious by a passerby about an hour after the race began. K.G. had experienced exertional heatstroke (EHS).

The local police and emergency medical technicians were summoned. Upon arrival, they incorrectly surmised that K.G. had suffered a head or spinal injury. After checking his airway, breathing, and circulation, the EMTs proceeded to secure his unconscious body onto a full-body board, to reduce the possibility of spinal cord injury and to allow transportation to a nearby hospital. Upon arrival at the emergency room, the medical staff did not take a rectal temperature because they were instructed that his problem was a head or neck injury. CT scans were performed at 12:45 P.M., 1:25 P.M., and 11:45 P.M. No injury or trauma was found in the lungs, thorax, head, or to discs in the spine. Only mild cerebral edema and hemorrhages could be found, and these supported a diagnosis of heatstroke.

K.G. experienced significant elevation of central body temperature (estimated to be 42°C, or 107.6°F, by an anesthesiologist) for a prolonged period of time. This is the most dangerous EHS scenario possible. Rapid recognition of EHS and whole-body cooling likely would have saved K.G.'s life. This should have been applied immediately after he was discovered unconscious. Cooling treatment for EHS is common knowledge at most mass participation events. In this case, confusion was established when the first person on the scene assumed that K.G. had fallen and suffered a head or spinal injury.

Hospital room records for the initial 2 hours of treatment indicated that the following conditions existed: liver failure, kidney failure, rhabdomyolysis, disseminated intravascular coagulation (DIC), coma, elevated liver enzymes (CPK to 148,000), ischemic bowel (leading to endotoxemia), and blood in the urine. These conditions are consistent with heatstroke and are not common in head or spinal cord injuries.

On the second day of treatment, K.G.'s deteriorating vital signs and laboratory test outcomes resulted in his transfer to a larger medical center, for more sophisticated treatment. Numerous specialists were consulted. K.G.'s condition, however, continued to decline. It is unlikely that any intervention at this point would have saved his life. The multisystem tissue damage due to hyperthermia was too much for his body to cope with.

The attending physicians attempted to maintain most body systems. They performed exploratory thoracic surgery, transfused blood to adjust blood electrolytes continuously, performed dialysis, inserted a pressure sensor to observe rapidly changing intracranial pressure, supported blood pressure with several medications, and even considered a liver transplant. Their fortitude constituted an outstanding comprehensive effort. They kept K.G. alive as long as possible. However, K.G. succumbed to multisystem organ damage.

It is difficult to identify why K.G. experienced EHS just 4.8 kilometers (3 miles) into the event. Perhaps he had a fever at the start of the race, caused by an unidentified viral or bacterial illness. As with many cases of EHS, the cause remains unknown.

Unfortunately, this race did not meet expected standards of race organization and safety. Prerace instructions and registration materials failed to provide sufficient information regarding the amount of water needed, possible alteration of race tactics to match the harsh environmental conditions, the course route, placement of water stations and mileage markers, and the risks involved with racing in hot–humid conditions. The organizers did not provide enough water, medical support, course maps, or course markings, and there were not enough course marshals, rescue and first-aid personnel, and medical assistants present. Finally, because the race course traversed dense woods, distress calls were difficult to recognize and respond to, and EMT access was hampered.

Exercise in the heat causes many homeostatic changes in the human body, including alterations in the circulatory, thermoregulatory, and endocrine systems. Many interrelated physiological processes work together to sustain central blood pressure, cool the body, maintain muscular function, and regulate fluid volume. Attempting to sustain exercise, especially intense activity, in a hot environment can overload the body's ability to properly respond to the imposed stress, resulting in hyperthermia, dehydration, deteriorated physical and mental performance, or a potentially fatal case of exertional heatstroke (EHS) (1, 2).

Motivated athletes, soldiers, and industrial laborers who exercise or work at a high intensity or for prolonged periods of time can experience an excessive rise in core body temperature, usually associated with dehydration (3). All athletes, coaches, and medical staff should understand EHS, including its pathophysiology, common signs and symptoms, treatment, and prevention. This chapter considers these issues.

PATHOPHYSIOLOGY

The causes and complications of the extreme hyperthermia associated with heatstroke are multifaceted. EHS differs from classical heatstroke and can be divided into three phases: acute, blood and enzyme, and late.

Classical Versus Exertional Heatstroke

To understand the etiology of heatstroke, one must understand its two variants: EHS and classical heatstroke (CHS). CHS occurs mostly in the elderly and usually in epidemics during summer heat waves in the northern parts of the United States (4). The dramatic hyperthermia that occurs in CHS is most likely to develop in individuals who have no air conditioning, little residential air circulation, and existing medical conditions that may not be adequately treated or that may require medications that inhibit normal thermoregulatory control (4, 5). The combination of these circumstances, especially in the presence of stifling environmental heat, can cause a rise in deep body temperature. CHS also commonly occurs when irresponsible guardians leave small children in vehicles or other small spaces in which unbearable heat accumulates.

Conversely, EHS usually occurs in isolated cases in young and healthy athletes, laborers, and soldiers who must exercise intensely in hot and humid conditions (4). EHS, by definition, is caused when the combined thermal stresses from the environment and muscular metabolism exceed the body's maximal heat dissipation via evaporation, radiation, convection, and conduction. If the scale is tipped dramatically in favor of heat storage, a dangerous hyperthermia ensues, and the condition of EHS may be the end product if steps are not taken to alter heat balance (3, 6, 7). The equation can be altered if the environmental heat stress or metabolic heat production are decreased or if heat loss is increased (1, 3, 8-10). Our concern is clearly with EHS; therefore, the remainder of this chapter focuses solely on EHS rather than on CHS.

Acute Phase of EHS

The acute phase of EHS is defined as the first hour after the initial collapse. EHS results from either overloading or failure of the thermoregulatory system in response to intense exercise, usually in a hot environment (3). The metabolic requirements of working muscle and cooling skin, exacerbated by temperature and humidity extremes, can overwhelm the body's capacity to dissipate heat. The body preferentially maintains arterial blood pressure over thermoregulation, skin dilation, and muscle perfusion (2). Ultimately, heat production exceeds heat dissipation, and core temperature rises dramatically, until dangerous hyperthermia exists (8, 9).

See figure 3.1 for an overview of the initial physiological responses that ultimately may cause a decrease in heat dissipation. Another potential mechanism not depicted in figure 3.1 is the possible overwhelming of the thermoregulatory system caused by the simple reality of heat acquisition exceeding maximal heat dissipation ability. The acute phase of EHS is characterized mostly by central nervous system (CNS) dysfunction (altered consciousness, convulsions, irrational behavior, coma, stupor, irritability, delirium, agitation, etc.) and distinct hyperthemia (11-15).

Blood and Enzyme Phase of EHS

The blood and enzyme phase of EHS occurs from 1 to 48 hours after initial collapse and is characterized by disorders of the blood constituents and enzyme function. Some of the altered enzymatic responses include elevated serum levels of alanine aminotransferase (ALT), aspartate aminotransferase (AST), lactate dehydrogenase (LDH), and creatine phosphokinase (CPK). These enzymes are elevated by damage to the organ systems, discussed later in the

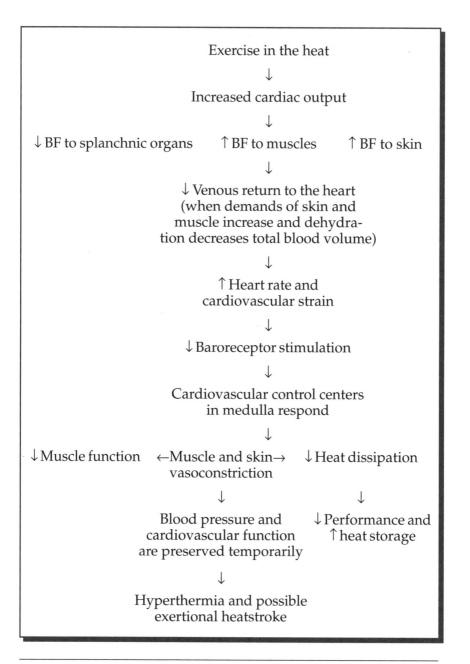

Exercise in the heat

↓

Increased cardiac output

↓

↓ BF to splanchnic organs   ↑ BF to muscles   ↑ BF to skin

↓

↓ Venous return to the heart
(when demands of skin and
muscle increase and dehydration decreases total blood volume)

↓

↑ Heart rate and
cardiovascular strain

↓

↓ Baroreceptor stimulation

↓

Cardiovascular control centers
in medulla respond

↓

↓ Muscle function   ←Muscle and skin→   ↓ Heat dissipation
vasoconstriction

↓   ↓

Blood pressure and   ↓ Performance and
cardiovascular function   ↑ heat storage
are preserved temporarily

↓

Hyperthermia and possible
exertional heatstroke

Figure 3.1 Potential circulatory responses to exercise in the heat. BF = Blood flow. ↑ = Increased. ↓ = Decreased.

Adapted, by permission, from D.J. Casa, 1999, "Exercise in the heat: Fundamentals of thermal physiology, performance implications, and dehydration," *Journal of Athletic Training* 34(3): 246-252.

chapter. The blood disorders may include endotoxemia (see the following sidebar), leukocytosis, elevated white blood cell (WBC) count, endothelial damage, clotting dysfunctions, hypophosphatemia, hyperkalemia (after a brief hypokalemia), and hypocalcemia (11-17).

Endotoxemia

A robust immune system is desirable at all times, particularly for athletes. However, the combination of overtraining and the stress of new surroundings suppresses immune function and often is blamed for illnesses (e.g., the common cold, sore throat, influenza, mononucleosis) that afflict athletes during physical training (18). Further, an immunological response to intestinal conditions may occur when an athlete experiences concurrent multiple stressors (19-22).

During digestion, **gram negative bacteria** exist in **chyme,** in the small and large intestines. Dead gram negative bacteria provide large amounts of the toxic cell wall component **lipopolysaccharide (LPS).** LPS found in the outer membrane of gram negative bacteria is known as **endotoxin.** A high level of plasma LPS seems to be the immediate cause of human septic shock (23). One of the most important discoveries in critical care medicine in the 1980s involved the recognition that LPS may leak out of damaged intestines into the blood, which in severe cases can result in cardiovascular insufficiency, extensive organ damage, or death.

When LPS enters the portal circulation, hypersecretion of **cytokines** (e.g., TNF, IL-1, IL-6) may occur; cytokines are toxic immune mediators that may cause fever, nausea, vomiting, diarrhea, headache, tissue injury, shock, or death. These symptoms are observed in many cases of heatstroke (24) and have led authorities to suggest that cytokine release occurs during exertional heatstroke (25). Although heatstroke often occurs without warning and has an unknown etiology in most cases, autopsies of human heatstroke victims have found high levels of plasma LPS and cytokines (26, 27). LPS also could be involved in heatstroke by suppressing sweating (28) or cardiac function (29).

Compared to exercise in cool environments, exercise–heat stress produces a markedly reduced blood flow in splanchnic vascular beds concurrent with an increased heart rate (30). This diversion of blood flow contributes to increased skin blood flow (important for heat dissipation) but carries the threat of compromising the function of splanchnic organs (31, 32). This is important because the removal of bacteria and other microorganisms is a normal function of the liver (29).

If exercise–heat stress or ischemia is great, an increase in plasma LPS may occur because of increased gut permeability. This phenomenon has been observed in primates, cats, swine, and rats (22, 33). Research with these animals demonstrated that core hyperthermia must reach severe levels (42-45°C; 107.6-113°F) before lethal increases in LPS occur (22, 25, 33, 34). Hypovolemia and diarrhea also increased the entry of gut-derived LPS into the circulation (22, 25).

In humans, elevated plasma LPS levels have been observed in triathletes and ultramarathon (89.5-km) competitors. For example, Bosenberg et al. (19) found that LPS rose and the "natural" anti-LPS IgG (i.e., the antibody formed in response to LPS) decreased during competition. Brock-Utne et al. (20) observed that 80% of collapsed runners had elevated levels of plasma LPS. In the casualties with low (normal) levels of LPS but high concentrations of anti-LPS IgG, symptoms were far less severe than in those casualties with high (abnormal) plasma LPS and low levels of anti-LPS IgG. This latter group required 2 days to recover, suggesting that higher levels of anti-LPS protected some runners.

Late Phase of EHS

The late phase of EHS starts at about 12 hours after initial collapse, overlapping with the blood and enzyme phase, and continues until 72 hours postcollapse. This phase is characterized primarily by liver and kidney dysfunction. Elevated ALT, AST, and bilirubin indicate hepatic failure and are likely precipitated by the effects of rhabdomyolysis (12, 14, 35, 36), a breakdown of skeletal muscle. Liver function often improves dramatically when other contributing factors are successfully treated. A liver transplant is rarely, if ever, needed. Acute renal failure may occur for many reasons, but sustained hypotension, direct thermal injury, and disseminated intravascular coagulation (DIC) are the most common causes. Additionally, the physiological milieu that exists during rhabdomyloysis (see the following sidebar), acidosis, or other pathological manifestations of EHS will influence the degree of renal dysfunction. These conditions are often represented by oliguria, anuria, thick blood (like "machine oil"), hematuria, and proteinuria (12, 13, 15).

Rhabdomyolysis and Acute Renal Failure

When strenuous, prolonged exercise is part of an EHS patient's history, exercise-induced skeletal muscle breakdown (i.e., rhabdomyolysis) and myoglobin in urine (i.e., myoglobinuria) are observed in 30 to 50% of the cases (37). Because other causes of rhabdomyolysis are known (e.g., burns, crush injuries, electrical shock), these conditions should be ruled out, along with myocardial infarction, stroke, and documented sepsis. Exercise-induced rhabdomyolysis may be complicated by elevated body temperature and shock. Most rhabdomyolysis patients present initially with confusion, pallor, and hyperthermia (38). These symptoms improve rapidly, only to have the patient rapidly deteriorate within the subsequent 12 to 24 hours with acute renal failure (ARF), which is a potentially lethal condition with a poor prognosis. Acute renal failure occurs in rhabdomyolysis when damaged muscle fibers release myoglobin (an oxygen-carrying protein). Myoglobin then precipitates in the nephrons of the kidneys and degrades into the component chemical ferrihemate, which is toxic to renal cells. Numerous complaints may arise during this period, including fever; neuronal or brain degeneration; and necrosis of the liver, skeletal muscle (e.g., pectoralis and tibialis anterior), and cardiac tissues. ARF occurs in approximately one out of every 10 to 25 EHS cases, depending on the duration and severity of hyperthermia, hypotension, dehydration, and myoglobinuria (37). Severe dehydration or extreme hyperthermia alone can lead to ARF.

The goal of treatment for rhabdomyolysis is to avoid life-threatening complications, especially ARF. Although the exact mechanism of ARF is not totally understood, the presence of ferrihemate (a dissociation product of myoglobin), which interferes with renal tubular function, and decreased renal blood flow secondary to dehydration are likely involved. When rhabdomyolysis is present, exercise should be discontinued for 3 to 7 days, the patient should properly hydrate, and the warning signs of ARF should be monitored. These include decreased urine output, darkening of urine color, heart palpitations, and irregular heart beat. The signs and symptoms of rhabdomyolysis and potential laboratory tests are presented in figure 3.2. For patients who require hospitalization, intravenous hydration should be implemented, to maintain urine output at 200 to 300 milliliters per hour.

In severe cases of rhabdomyolysis, the return to activity should be controlled and clinical laboratory tests (figure 3.2) should be repeated. When CK levels have returned to the

Signs and symptoms

Myoglobinuria	Dark or red urine
Decreased urine volume	Elevated serum creatine
Muscle weakness	phosphokinase
Muscle soreness, swelling, and tenderness during palpation and passive stretching	Decreased range of motion (active and passive)

Laboratory tests to consider

Creatine phosphokinase (CPK)	Serum and urine myoglobin
Alanine aminotransferase (ALT)	Aspartate aminotransferase (AST)
Potassium	Calcium
Albumin	Phosphorus
Lactate dehydrogenase (LDH)	Blood urea nitrogen (BUN)
Creatinine	

Figure 3.2 Clinical presentation of rhabdomyolysis and laboratory tests. Data from 37, 39, 40.

normal range, the patient may resume a moderate exercise program, although return to full activities may require weeks. The reader may consult Armstrong, DeLuca, and Hubbard (41) for a detailed account of recovery in 10 former EHS patients, all of whom were male soldiers. All but one of these men (subject A) recovered fully by 61 days, as evidenced by normal exercise–heat tolerance, and had normal renal responses. The lone patient exhibited renal concentrating dysfunction (i.e., consistent urine specific gravities of 1.001 to 1.005), with normal exercise–heat tolerance, up to 11.5 months after EHS. A comparison of hospital records indicated that the mean CPK level for recovered soldiers peaked 1 to 2 days postepisode at 4,108 units per liter, whereas subject A experienced a peak CPK of 18,120 units per liter 5 days after EHS, indicating a prolonged time course with extensive muscle degradation. The investigators recommended that all severe EHS patients be tested before return to full activities, by evaluating their thermoregulation, sweat gland function, whole-body electrolyte balance, blood constituents, and ability to acclimate to exercise in the heat over 7 to 14 days.

Hyperthermic Organ Dysfunction

It seems that none of the organ systems are spared from the potential consequences associated with severe cases of EHS. Table 3.1 provides an overview of the organs that may be affected by EHS.

Table 3.1
Potential Organ Damage Associated With Exertional Heatstroke (EHS)

Physiological system	Possible organ impairments during EHS
Central nervous system	Cerebral edema, petechial hemorrhage, degeneration of nerve cells (especially in the cerebellum), cerebral hemorrhage (in response to severe coagulopathies), coma, convulsions, delirium, oculogyric crisis, cerebellar syndromes
Liver	Jaundice, leakage of liver enzymes (ALT, AST) into blood (indicating heptocellular damage), elevated alkaline phosphatase (indicating impairment of liver's excretory function), centrilobular necrosis, perisinusoidal edema, impaired synthetic function, decreases in serum albumin, decreases in coagulation factors, bilirubin
Kidney	Acute renal failure, proteinuria, hematuria, DIC, tubular damage from myoglobinuria, hyperuricaemia, hypokalemia, olguria, anuria, acute tubular necrosis, interstitial nephritis
Blood	Reduced coagulation ability, reduced clotting factors, DIC (a coagulopathy involving factor VIII and V), consumption coagulopathy, thrombocytopenia, elevated WBC count, hemorrhage (possibly into skin, lungs, heart, GI tract), fibrinolysis, endotoxemia
Heart	Sinus tachycardia, arterial hypotension, elevated central venous pressure, acute left heart failure, conduction disturbances, hemorrhages, fragmentation and ruptured muscle fibers noted with edema
Lungs	Tachypnea, alveolar hyperventilation, pulmonary edema usually associated with adult respiratory distress syndrome, arterial hypoxemia, hemorrhages
GI tract	Nausea, vomiting, diarrhea, bleeding, necrosis of small intestine

Physiological system	Possible organ impairments during EHS
Electrolytes	Hyperkalemia (hypokalemia initially until rhabdomyolysis or kidney failure take effect), hypocalcemia (calcium salts deposit in damaged muscle), hypophosphatemia (from respiratory alkalosis) or hyperphosphatemia (if released from muscle)
Acid–base balance	Respiratory alkalosis and lactic acidosis initially, but metabolic acidosis is usually the dominant feature due to renal failure and muscle damage
Muscle	Rhabdomyolysis (damage to the muscle fibers causes pain, weakness, and efflux of myoglobin, potassium, enzymes, and uric acid into the bloodstream), elevated CPK indicating muscle damage

ALT = alanine aminotransferase; AST = aspartate aminotransferase; CPK = creatine phosphokinase; DIC = disseminated intravascular coagulation; GI = gastrointestinal; WBC = white blood cell.
Data from 10-12, 14, 15, 42, 43.

EPIDEMIOLOGY AND PROGNOSIS

Accurately assessing the incidence of EHS is nearly impossible. Much of the available research provides only the number of deaths from EHS, not the actual number of cases. Additionally, the incidence data could be misleading because an event, such as a fun run or military training maneuver, that shows zero risk for a number of consecutive years may have a large incidence if the event occurs during a heat wave. Many cases go unreported because the deaths are attributed to the organ systems that were the end cause of death rather than to the original cause, which may have been EHS. In addition, many physicians still diagnose cases of EHS as heat exhaustion (see case report 3.2), further decreasing incidence numbers. EHS, unlike many medical conditions, is largely preventable. However, although the precautions and medical care instituted at a venue can reduce the incidence of EHS and mortality rates, the risk of EHS can never be eliminated. With that said, the available data indicates that during athletic and military activity that requires intense exercise in the heat, the incidence of EHS is approximately 1 out of every 1,000 participants (44, 45). This number can be dramatically altered (up or down) depending on the precautions taken by the medical staff, the environmental conditions, the intensity of exercise, and numerous other factors mentioned in this chapter. Improved knowledge of the identification and treatment of EHS has increased the survival rate of EHS patients.

When a person with EHS receives prompt and appropriate medical diagnosis and treatment (i.e., rapid cooling, arresting of convulsions, rehydration), the survival rate has been reported to be greater than 90 to 95% (11, 14, 46). Inappropriate medical attention dramatically reduces the likelihood of full recovery. The most important factors influencing prognosis are the degree of hyperthermia and the length of time the individual is above a critical hyperthermia (about 40-41°C; 104-105.8°F) (11, 14, 46, 47). Other predictors of prognosis include the duration of coma, hyperkalemia, oliguric renal failure, and levels of serum enzymes (e.g., ALT, AST, and LDH). A well-trained medical professional who quickly identifies the presence of EHS, promptly initiates rapid body cooling, and quickly evacuates the patient to a hospital maximizes the possibility of a healthy outcome. Proper treatment at the hospital, including continued cooling if necessary, and restoration of fluids also optimize the prognosis. Monitoring organ function within the first 24 to 48 hours after insult is critical, because the patient's condition may rapidly worsen; prompt and appropriate care will unquestionably enhance the prognosis.

IDENTIFICATION OF EXERTIONAL HEATSTROKE

EHS is a medical emergency and should be treated as such. Immediate recognition of symptoms and initiation of treatment are necessary to maximize the potential for a complete restoration of normal physiological function. Negligence on the part of supervisors or medical staff can result in potentially fatal consequences.

The two critical variables for a diagnosis of EHS are thermoregulatory failure (i.e., elevated core temperature) and obvious CNS impairment (i.e., altered consciousness) (3, 16, 17). Core temperature in patients with EHS is higher than 40°C (104°F) (see the following sidebar). Other common signs and symptoms may include hypotension, tachycardia, increased respiratory rate, vomiting, diarrhea, dehydration, coma, and convulsions (13, 16, 47). Blood analyses may reveal elevated serum enzymes (e.g., ALT, AST, and LDH) (13, 16, 47). Sweating will most likely be present at the onset of EHS, and dehydration is likely but not always present (13, 16, 47). Many misconceptions associated with this condition are presented in table 3.2.

Assessing Core Temperature

Measuring core temperature is critical for diagnosing EHS (13, 16). It is difficult to discriminate between EHS and exertional heat exhaustion (EHE) during triage because they share many signs and symptoms (3, 47). The medical staff should consider central nervous system (CNS) impairment

Table 3.2

Common Misconceptions Associated With
Exertional Heatstroke (EHS)

Misconception	Reality
Athlete is no longer sweating when he or she collapses.	In most cases, an athlete will still be sweating when collapse occurs. The medical team should not use the absence of dry skin as a key factor in diagnosing exertional heat exhaustion (EHE) instead of EHS.
Rectal temperature must be greater than 40°C (104°F).	In most cases, the first valid rectal temperature measurement is ascertained upon arrival at the hospital, 15 to 60 minutes after collapse. The core temperature at this time might not reveal the peak of hyperthermia.
Athlete must be severely dehydrated.	While it is likely that an athlete will have experienced substantial dehydration, EHS may occur in as little as 20 to 30 minutes if exercise is intense and environmental conditions are severe. In this case, fluid loss may be minor.
Athlete can only succumb from EHS after a lengthy exercise session.	Most cases of EHS occur during the first 2 hours of exercise. The longer the activity, the lower the exercise intensity, resulting in a decreased likelihood of overwhelming the thermoregulatory system. Short duration activities may involve higher exercise intensity. High-intensity exercise is the number one factor contributing to EHS.
Athlete will only succumb to EHS on hot and humid days.	Although the incidence of EHS is directly correlated with the wet bulb globe temperature (WBGT), many cases of EHS occur on mild days when the medical staff considers the risk to be very low. Risk factors such as febrile illness, obesity, carrying equipment, low fitness, lack of acclimation, very intense exercise, no rest period, and significant dehydration will increase the likelihood of an episode in mild environmental conditions (see chapter 8).
Cooling the athlete with cold or ice water immersion causes peripheral vasoconstriction and inhibits heat dissipation.	Reducing the amount of time that an athlete experiences hyperthermia is the most critical factor in the morbidity and mortality associated with EHS. Water is, without question, the best medium to maximize heat loss. The risks associated with this type of cooling are unfounded and not significant.

(continued)

Table 3.2

(continued)

Misconception	Reality
All athletes who succumb to EHS will also become unconscious.	Athletes who succumb to EHS will, by definition, have central nervous system disturbances but may not necessarily become unconscious. They may have lucid periods, exhibit strange or inappropriate behaviors, say things that do not make sense, or not be able to thoughtfully process questions that the medical staff poses.
Medical staff should focus on cooling first and not be concerned with fluid replacement.	While cooling is absolutely critical for thermoregulatory and cardiovascular reasons, an IV line should be placed as soon as possible in a case of EHS. The infusion of 1 to 1.5 liters of 1/2 normal saline in the first 30 to 45 minutes assists the recovery of cardiovascular function and numerous other physiological functions. This modest amount of fluid will not put the athlete at risk of pulmonary edema. Further fluid replacement should be based upon laboratory values, history, and signs and symptoms.
EHS arises gradually, and athletes exhibit specific signs and symptoms prior to collapse.	In many cases, athletes do not experience, and medical staff do not notice, prodromal symptoms. Teammates, coaches, and medical staff should be aware that this condition can appear quite suddenly.
Predicting core temperature with an infrared scanner provides an estimate of the degree of hyperthermia.	Medical staff should never utilize tympanic temperatures to diagnose EHS. These measures have been found to be grossly inaccurate when compared to rectal temperatures in hyperthermic athletes. In one study, tympanic temperature averaged 2.3°C (4.1°F) lower than rectal temperature. See the sidebar on page 40.

Data from 3, 6, 9-17, 42, 44-61.

(e.g., unconscious, altered thought process, convulsions, combative or inappropriate actions) and hyperthermia as the two most revealing clues indicating EHS. When assessing core temperature, the medical staff should never, under any circumstances, rely on oral, axillary, or tympanic temperature readings in a person who has been exercising to determine the presence of EHS. These methods of predicting core body temperature are not valid in hyperthermic individuals during or immediately after exercise. Medical staff should use rectal temperature readings.

One recent study found tympanic temperature measurements to be, on average, 2.3°C (4.1°F) lower than rectal temperatures in hyperthermic runners (54). Another study revealed that temperatures were, on average, 2.1°C (3.8°F) lower via a tympanic measure as compared to rectal temperature in heatstroke patients (45). Although esophageal temperature is the most reflective of the true core temperature at any given moment, rectal temperature is by far the most practical field measure of core temperature that offers a valid indication of the magnitude of hyperthermia. Athletic trainers, emergency medical technicians, and physicians should have equipment available and be trained to ascertain rectal temperatures on-site. A regular rectal thermometer suffices, but a rectal thermistor is preferable because it can be kept in place to allow perpetual monitoring if cold or ice water immersion must be administered to rapidly cool the individual.

In the absence of the availability of a valid core temperature measurement, the medical staff on-site should rely on other critical signs and symptoms to differentiate between EHS and EHE. The critical variable is CNS dysfunction, especially alterations in consciousness. Relying on invalid measures of core temperature in hyperthermic individuals (e.g., tympanic temperature) may cause fatal consequences if appropriate treatment and medical attention is delayed. Additionally, because athletic trainers are usually the medical staff that identifies the presence of EHS in athletic situations, they should be adequately equipped and trained to assess core temperature via rectal thermometers or rectal thermistors.

The signs and symptoms before and during EHS, along with treatment protocols, are shown in table 3.3.

Other conditions that may cause a person to collapse or have CNS dysfunction are listed in figure 3.3. It is vital that the medical staff determine immediately if the patient has EHS or exertional heat exhaustion (EHE). Figure 3.4 offers five critical questions that can help simplify this process. A "yes" answer to any of these questions should raise the suspicion of EHS. The process of discriminating between EHS and hyponatremia is discussed in chapter 9.

Table 3.3

Prevention, Identification, and Treatment of Exertional Heatstroke (EHS

Prevention		Identification
Preexercise	**During exercise**	**Precollapse**
Consider WBGT*: If WBGT is high, consider operating under high alert, decreasing exercise intensity, increasing the length and/or number of rest breaks, or rescheduling or canceling the event. Perhaps modify activity for high-risk individuals.	Be flexible: Environmental conditions may worsen or lessen; this may warrant a change in protocol.	Irrational behavior, disorientation, inappropriate comments, irritability.
Plan rest for cooling and rehydration.	During rest breaks (in the shade if possible), mandate and monitor rehydration.	Confusion, disorientation, altered consciousness.
Talk to participants, coaches, officials, and medical staff about risks, signs and symptoms, and treatment of EHS.	Monitor hydration status during prolonged exercise or between extended rest periods (body weight is a simple and convenient method).	Increased body temperature by measurement or report of feeling excessively warm.
Consider which individuals are at highest risk for EHS (see chapter 8).	Watch for decreases in performance and early signs and symptoms of dehydration and EHS.	Increasing dehydration.
Ensure that individuals are properly conditioned and heat acclimated.	Remind participants, medical staff, and coaches to inform you if they see any peculiar behavior.	Nausea or vomiting. Diarrhea.
Ensure that individuals are properly hydrated.	Make sure individuals have proper clothes and minimal equipment during extreme environmental conditions.	Profuse sweating. Decreasing performance, loss of desire, or weakness.
Know individuals' baseline body weights.	Talk with participants during rest periods to determine health status, fluid needs, and level of warmth.	Sluggish feeling, general malaise, or drowsiness.
Have proper fluids available.	Consider using cold towels, ice packs, or other methods of cooling during rest periods.	Headache.
Have proper staff available to monitor hydration and medical conditions.	Pay special attention during very intense activity and during a second daily practice that occurs in the middle of the afternoon.	Dizziness. Dry skin.
Have proper medical equipment and communication devices available on-site to identify and treat EHS (see chapter 9).	Monitor high-risk individuals (those with low fitness, recent or current illness, lack of acclimation, obesity, drug or alcohol use, history of EHS).	
Notify local EMTs and hospitals of potential cases (especially when working large-scale events).	Continue to monitor WBGT.	
Regularly review the ways to prevent, identify, and treat EHS.	Check fluid, ice, and equipment to be sure proper supplies are available.	

Note that not all of the prevention steps, identification signs and symptoms, and treatment techniques are appropriate or reported in all cases of EHS. This is intended to be a thorough list of the possible prevention and treatment steps and signs and symptoms that may be found upon examination. See text and chapter 9 for additional information.

	Treatment	
Immediate postcollapse	**Within 1 hour postcollapse**	**1 to 48 hours postcollapse**
Coma (sustained loss of consciousness).	Cool the person as quickly as possible via ice or cold water immersion in shade. Use alternate means if this is not possible (e.g., ice bags, air-conditioning, cold towels, fanning, wetting the skin with cold water).	Monitor for brain pathologies including edema, congestion, and hemorrhages.
Confusion, irrational behavior, or agitation.		Monitor for blood pathologies, especially disseminated intravascular coagulation.
Elevated (greater than about 40°C, 104°F) core temperature when measured rectally (do not use tympanic measure).	Monitor core temperature via rectal temperature and stop cooling at about 38-39°C (100.4-102.2°F).	Monitor for liver pathologies, including centrilobular necrosis.
	Activate emergency medical system and transport to hospital, unless appropriate staff and equipment are on-site or case is minor and patient responds to treatment.	Monitor for muscle pathologies, especially rhabdomyolysis and lactic acidosis.
Convulsions or seizures.		
Hot and dry skin (though wet skin is found upon initial collapse in most cases of EHS).	Provide intravenous fluid (be sure not to give beyond fluid losses) if oral fluid is not possible or if medications (e.g., anticonvulsive drugs) are required.	Monitor for kidney pathologies, especially acute tubular necrosis.
Hypotension (systolic pressure likely lower than 100 mmHg).	Conduct an electrocardiogram to assess cardiovascular function. Treat positive findings.	Monitor for gastrointestinal tract pathologies, especially general bleeding diathesis and ulcerations.
Sustained increase in respiratory rate.	Obtain acute enzyme readings (alanine aminotransferase, aspartate aminotransferase, creatine kinase, lactate dehydrogenase) to ascertain muscle, liver, and kidney function. Treat positive findings.	Monitor for lung pathologies, including respiratory alkalosis, aspiration pneumonia, and hemorrhagic pneumonia.
Sustained tachycardia, when heart rate normally decreases after cessation of exercise.	Check serum urea, electrolytes, glucose, calcium, phosphate, osmolality, creatinine, hemoglobin, white blood cells, platelets, and pH.	Monitor for heart pathologies, including myocardial necrosis and hemorrhagic myocarditis.
Dehydration.	Check urine for protein, myoglobin, casts, osmolality, and volume. Treat positive findings.	Monitor for hypo- or hyperhydration, because fluid deficits may not yet be replenished or may have been excessively replaced.
Vomiting and/or diarrhea.	Monitor vital signs (respiratory rate, heart rate, blood pressure) and the decreasing or increasing of any other signs and symptoms.	
Decrease in sweating rate in cases of prolonged central nervous system involvement.	Maintain airway, because aspiration is common.	Monitor core temperature via rectal or esophageal thermometer.
Rapid onset of signs and symptoms in exercising individuals.	Consider supplemental oxygen administration, and use positive pressure if this does not suffice.	

*See the ACSM Position Stand on Heat and Cold Illnesses During Distance Running, summarized in appendix A, for information on wet bulb globe temperature (WBGT).
Data from 3, 6, 7, 10-17, 41, 43-48, 50-59, 62-71.

Dehydration	Animal bite or sting (bee, ants, snake)
Heat exhaustion	
Malignant hyperthermia	Drug intoxication
Hyponatremia	Animal poisoning
Heat syncope	Cerebral injury (e.g., concussion, hematoma)
Encephalitis	
Meningitis	Hypothalamic hemorrhage
Epilepsy	Coagulopathies
Hypoglycemia	Cerebrovascular accident

Figure 3.3 Conditions other than exertional heatstroke that may cause individuals to collapse or have central nervous system dysfunction.
Data from 13, 14, 16, 17, 46.

1. Does the person have an altered level of consciousness (unconscious, semiconscious, trouble understanding or answering questions, irrational behavior or comments, stupor, agitation, delirium)?

2. Does the person have significant hyperthermia (rectal temperature greater than 40°C, 104°F)?

3. Is the person suffering from convulsions?

4. Has the person collapsed during intense exercise in hot conditions?

5. After the initial collapse, is the person's skin dry or relatively dry when copious sweat is obvious on the skin of other individuals? Remember that most people suffering from EHS are still sweating at the time of collapse.

Note: A "yes" answer to *any* of these questions should raise the level of awareness, and EHS should be suspected. A "yes" to just one of these questions may indicate EHS. A "yes" to both numbers 1 and 2 should conclusively indicate EHS.

Figure 3.4 Five questions to determine if a collapsed person has exertional heatstroke or exertional heat exhaustion (EHE).

MEDICAL TREATMENT OF EXERTIONAL HEATSTROKE

The critical moment for the treatment of EHS is at the time of collapse. Herculean efforts should be made to rapidly cool the individual and then, after the body has become normothermic, to monitor for potential complications.

Acute Phase: First Hour (aka, The Golden Hour)

During the first hour after collapse, after airway, breathing, and circulation have been inspected, whole-body cooling is the primary concern (figure 3.5). The most effective method for the immediate treatment of EHS, because of its superior whole-body cooling rates (48) (figure 3.6) and lowest mortality rates, is cold or ice water immersion (approximately 5-15°C; 41-59°F) (47, 48, 52, 53, 57, 60, 70). A quick cooling rate is critical to survival; minimizing the time the organ systems are hyperthermic will maximize survival rates (47). If available equipment does not allow for water immersion, ice packs should be placed on the victim's head, neck, axillary, proximal femurs, and behind the knees (3, 11). Fans also will assist in cooling (43). Cooling should be continued until rectal temperature reaches approximately 38-39°C (100.4-102°F). An overview of treatment for EHS is covered in detail in table 3.3 and figure 3.5.

Three unfounded concerns associated with cold and ice water immersion include (1) hypothermic overshoot, (2) peripheral vasoconstriction, and (3) cardiogenic shock (11, 59, 60). The authors' field experiences and the scientific literature do not support these concerns as valid reasons to avoid water immersion (44, 47, 48, 52, 53, 57). Minimizing the duration of hyperthermia is the most critical variable in avoiding the possible negative sequelae of organ failure and death (47, 48).

While hypothermic overshoot is certainly a possibility when cooling via cold and ice water immersion, it is a calculated risk associated with treatment. The risk of having to treat a mild case of hypothermia after severe hyperthermia is acceptable and necessary. Constant monitoring of core temperature via a rectal thermistor, understanding and calculating cooling rates, and prompt removal from the immersion bath when rectal temperature reaches 38-39°C (100.4-102.2°F) will help to minimize the occurrence and severity of hypothermic overshoot.

The possibility of peripheral vasoconstriction exists, but water is an outstanding medium to maximize heat transfer. Its effects greatly supersede any potential consequences of peripheral vasoconstriction. One is reminded of the unfortunate fate of sailors plunged from capsized ships into frigid ocean waters (49). When these sailors

1. Have an assistant begin to fill a child's wading pool with ice and water in a nearby shaded area (or already have half-filled with water and add ice from ice chests kept next to the wading pool).
2. Check and record airway, breathing, circulation, and blood pressure.
3. Evaluate for central nervous system dysfunction.
4. Determine core body temperature with a rectal thermometer or thermistor.
5. Begin to aggressively cool the individual in cold or ice water if results of numbers 3 and 4 are positive.
6. Contact emergency medical system; inform them of a suspected exertional heatstroke so that they can be prepared in the ambulance and at the hospital.
7. Look for other signs and symptoms (if the individual is unconscious, talk to teammates/coworkers for additional information).
8. If appropriate medical staff is available, begin an intravenous line and provide 0.45% NaCl intravenous fluid at a rate of about 1 to 2 liters per hour (based on fluid needs).
9. Continue to monitor rectal temperature and remove from bath when the individual reaches 38-39°C (100.4°F) or when ambulance arrives (if the athlete is still hyperthermic, continue to cool the athlete in the ambulance via ice bags on peripheral arteries and cold towels on the head, chest, feet, and hands). Consider continued cooling before transport if appropriate medical staff is on-site.
10. If possible, escort the individual to the hospital to properly communicate all findings and to continue aggressive treatment, if necessary.

Figure 3.5 Ten immediate steps to take in the treatment of exertional heatstroke.

succumb to the frigid waters, they invariably die not from overheating but from hypothermia because of the rapid heat transfer that occurs when immersed in cold water. The same physiological response that has such unfortunate consequences for sailors can be used advantageously for hyperthermic individuals. That is, the rate of heat transfer in the medium of water is physiologically superior to that of air, and the immersion maximizes convective and conduc-

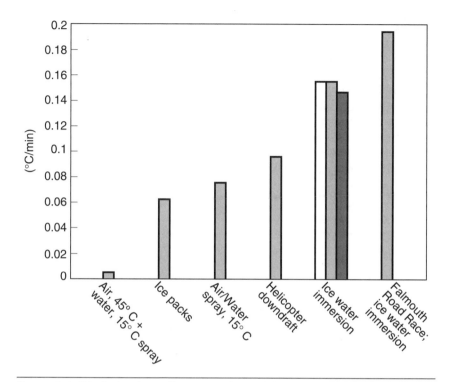

Figure 3.6 Whole-body cooling rates: Six modalities.
Data from 43.

tive cooling while allowing some evaporation. As for the third concern, cardiogenic shock, this condition has not been found to be an issue during ice/cold water immersion treatment of EHS.

If possible, an intravenous line should be placed in an antecubital vein of one of the arms that is hanging outside the tub. A minimum of 1 to1.5 liters of 0.45% NaCl should be infused. The fluid volume should be based on the degree of dehydration. This immediate restoration of some of the fluid deficit helps to reestablish normal cardiovascular function (16, 17). The intravenous fluid also may contain simple sugars or anticonvulsive therapies if the individual is experiencing hypoglycemia or convulsions, respectively (11, 13).

After One Hour

Table 3.3 provides an overview of pathologies that must be considered and treated if present within the first 48 to 72 hours after an episode of EHS. Disruptions of CNS function usually present as convulsion, comas, and disturbed behavior. Convulsions that do not respond immediately to diazepam may require phenytoin, with the caveat of an increased risk of arrhythmias. Brain edema can often be prevented by replacing fluid based on individual fluid needs rather than published recommendations. Pathologies of skeletal muscle are likely, and signs include acute renal failure, hyperkalemia, hyperuricaemia, hypocalcemia, hyperphosphatemia, and enzyme release. Hyperkalemia, a common and potentially dangerous consequence of muscle (e.g., rhabdomyolysis) and renal damage associated with EHS, can be treated with IV calcium, insulin, and glucose, followed soon after by ion exchange resins. Additionally, proper ventilation, oxygen, and volume repletion assist with hyperkalemia. In the most severe cases of rhabdomyolysis, dialysis may be necessary (11-15, 43).

The blood pathologies will usually present with fragmentation of red blood cells, hemorrhage, thrombosis, hemolysis, or thromboctopenia. Treatment of the DIC that causes these blood disturbances includes limiting muscle damage via proper cooling. If significant muscle damage has occurred, the ensuing hemorrhage can be treated by transfusion of fresh frozen plasma, cryoprecipitate, or platelet concentrates. Heparin and fibrinolytic therapy is not recommended. Coagulation tests should be done every 12 hours for the first 2 to 3 days, because DIC can arise quickly and has a dramatically enhanced recovery if identified promptly (11-15, 43).

Kidney disorders may result in acute renal failure as demonstrated by oliguria, acidosis, or hyperkalemia. Hyperkalemia can be treated as mentioned previously, and renal failure can be treated conservatively with dopamine infusion or more aggressively with mannitol or furosemide to stimulate diuresis. These last two should only be considered if volume repletion has been adequate. A final option would be to consider peritoneal or hemodialysis (11-15, 43).

Liver pathologies may lead to liver cell failure characterized by jaundice, hemorrhage, hypoglycemia, or enzyme release. Pathologies of the lung may result in tetany or hypoxia. Pathologies of the heart may lead to shock. This can usually be averted if fluid replacement closely approximates fluid losses, cooling is successful, and supplemental oxygen is provided. Pathologies of the gastrointestinal tract will likely result in nausea, vomiting, diarrhea, or hemorrhage (11-15, 43). Additional information is provided in chapter 9.

Prompt cooling and restoration of fluid losses encourage normal perfusion to the organ systems and can prevent the complications mentioned. Adequate perfusion allows proper oxygen availability and removal of waste products and heat, thereby limiting the potential damage and maximizing the likelihood of an uneventful recovery from EHS (11-13, 16, 47). Clearly, rapid and accurate anticipation, identification, and treatment of EHS improve the chances of complete recovery. When EHS is cared for promptly, the potential catastrophic sequelae in organ systems are completely avoided or dramatically minimized.

CONSIDERATIONS FOR RETURN TO TRAINING

Although no formal guidelines regarding returning to sport after EHS have been published, we recommend the following general guidelines for individuals who wish to return to training after EHS. These recommendations are based on the current knowledge base regarding recovery from EHS (1, 44, 61, 67, 72-74). First, we recommend that the patient refrain from exercise for 7 days after release from the hospital (not 7 days from the EHS episode, because serious cases require longer hospital stays). During these 7 days, we recommend that the athlete self-monitor his or her physical condition and immediately contact a physician if any signs or symptoms persist or appear for the first time.

Second, we recommend that the individual visit a physician approximately 7 days after leaving the hospital. At this time the physician should clear the individual for future participation if laboratory tests and examination indicate recovery. We strongly recommend that a heat-tolerance test be administered about 3 to 4 weeks after the incident to evaluate responses to exercise in the heat.

Third, after the physician's clearance, we recommend a gradual 7- to 14-day increase in the intensity and duration of exercise during the return to normal training patterns. Additionally, this 7 to 14 days will allow the reestablishment of heat acclimation and tolerance. Fourth, the physician should clear the individual for full participation if the 7 to 14 days of training prove fruitful and uneventful. Finally, we recommend that the supervising medical staff (e.g., athletic trainer or physician) investigate why the EHS occurred and take precautions to limit the likelihood of a future episode. Instituting proper event policies and individual participation guidelines can reduce the risk of future cases of EHS.

PREVENTING EXERTIONAL HEATSTROKE BY UNDERSTANDING THE MOST COMMON SCENARIOS

When reviewing the literature, especially regarding sport participation, a few scenarios present themselves with startling regularity (5, 6, 9, 11, 13, 14, 17, 42, 44, 46-48, 50, 51, 54-56, 71). In nearly all cases, EHS could have been prevented or identified earlier, thereby limiting the degree of morbidity and mortality.

Preventing EHS

Coaches, parents, officials, athletes, administrators, and medical staff all share in the responsibility of minimizing the likelihood of EHS. *Coaches* can schedule practices and events at a time of day when the environmental conditions are not extreme, modify the length and frequency of rest breaks based on the conditions, formally encourage rehydration, encourage healthy nutritional habits away from the practice field, and plan gradual onset of exercise in the heat to maximize acclimation. *Parents* can encourage proper hydration and inform the coach and medical staff when their children have febrile disease or another predisposing condition that may increase the risk of sustaining EHS. *Officials* can modify the game rules (especially in youth events) to allow unscheduled water and rest breaks, encourage game time changes when oppressive conditions exist, and watch

Both athletes and medical personnel have a responsibility in treating and preventing EHS.

players for unusual behavior during the event. *Athletes* can empower themselves by rehydrating according to fluid needs, being honest with medical staff regarding health conditions during practices and competitions, attaining a high level of fitness and acclimation before the beginning of practices, and watching their teammates for warning signs of EHS. *Administrators* can work with coaches to schedule events, field availability, and support services at a time of day when the environmental stresses may be minimized. In cases when an event must be scheduled during a high-risk time of the day, administrators can ensure that the event has ample medical staff, equipment, supplies, and support staff to provide a safe experience and appropriate and prompt care should a situation arise.

The greatest burden of preventing EHS lies with the *medical staff*. For athletic participation in the United States at the high school, collegiate, and professional levels, the certified athletic trainer (ATC) usually assumes the responsibility of prevention. The athletic trainer's role in preventing, identifying, and treating EHS is of profound importance because in nearly all cases the ATC is the medical staff that is available before, during, and after all practices, conditioning sessions, and competitions. For this reason the ATC must be proactive in identifying individuals who are at the highest risk (see the subsequent discussion and chapter 8) so that special attention can be given to these individuals. Additionally, the ATC can implement the pre-event preventive strategies noted in table 3.3 to maximize the likelihood of a safe and productive athletic experience.

Most Common Scenarios of Onset

Chapter 8 outlines some of the most common risk factors of sustaining an exertional heat illness. EHS is likely to occur when multiple risk factors exist. Many EHS patients exhibit clues of impending illness before the exercise session and these clues become more apparent during activity. A keen, suspecting, and proactive sense of awareness can save a life.

One scenario that seems to arise time and again involves a driven, overweight, out-of-shape, unacclimated football lineman. This athlete collapses at the end of the second practice of the day in the middle of the afternoon while doing highly intense activity; the athlete is dehydrated from previous training sessions and is not acclimated to the current heat wave. This scenario occurs during many high school and college football practices across the United States in the month of August. Fortunately, such collapses are rare, but in this and other situations, the risk is ever present and medical professionals can play an important role in preventing EHS. Early and rapid response can avert a tragedy.

Exertional Heatstroke in a High School Runner

Case Report 3.2

J.D., a healthy, 16-year-old, 170-centimeter (5-ft, 7-in.), 54.5-kilogram (120-lb) white male, collapsed with less than 200 meters to go in a 10,000-meter track race. This race took place in the first week of August during a heat wave in upstate New York. The temperature at the end of the race (10:15 A.M.) was 30.5°C (87°F) and the relative humidity was 70%. To complicate matters, the track's surface was extremely hot because of its black surface, no wind was blowing at the track, and no fluids were distributed during the race. J.D. was fully heat acclimated, had no history of heat illnesses, and was at peak fitness. He had no known illness at the time of race and was slightly dehydrated (about 1-2%) at the start. After 24.5 laps (9,800 meters), J.D. collapsed. Before collapse, J.D. recalled feeling extremely thirsty, an intense muscular fatigue, profuse sweating, and "suffocating warmth." He quickly got back up and collapsed again about 100 meters later. At this point he tried to stand, but his physical condition and the prompt actions of the medical staff precluded this action. J.D. was unconscious within 20 to 30 seconds after the second collapse and did not regain normal consciousness for about 2.5 hours. The medical staff at the track noted profuse sweating. J.D. was immediately transported to an ambulance, where ice packs were placed over superficial arteries and cold towels were placed on his chest and head. At this point, the athlete was unconscious with brief lucid moments, and this altered state of consciousness continued for the aforementioned 2.5 hours.

Upon arrival at the hospital, J.D. immediately was given an IV and placed on a bed of ice with a cooling blanket over him. An intravenous line was inserted and a 0.45% NaCl solution was administered; about 2.5 liters were given in the first hour. Minor convulsions were noted upon collapse and in the ambulance, but not in the hospital. The first rectal temperature measurement (40.2°C; 104.4°F) was taken approximately 20 to 25 minutes after collapse. Considering normal cooling rates and the immediate and appropriate treatment administered by the athletic trainers on-site, the emergency medical technicians during transport, and the emergency room staff at the hospital, it is reasonable to conclude that J.D.'s rectal temperature at collapse was approximately 41.4 to 42.2°C (106.5-108°F).

The medical staff stated that upon arrival at the hospital, J.D. was unable to speak coherently except regarding his father's work phone number. In the first 2 hours after arrival in the hospital, this phone number was the only coherent verbal response the athlete offered. It is interesting that the athlete later reported having heard many questions that the doctors and nurses asked during this time and remembers saying the answers (at least he thinks he said the responses), but these responses were never understood by the medical staff.

Also, upon admission J.D. was tachycardic (pulse of 144 beats/min), had a high respiratory rate (26-27 breaths/min), and had a blood pressure of 120/75 mmHg. Other serum measurements taken upon arrival, including sodium, potassium, chloride, hemoglobin, and white blood cells, were normal. Additionally, an electrocardiogram upon arrival indicated a "borderline normal ECG or normal variant" and a "sinus arrhythmia." Total cooling time was approximately 60 minutes and ceased when J.D.'s temperature reached 38°C (100.4°F). After about 2 hours in the hospital, J.D. slowly regained consciousness and began to verbally communicate with the doctors and nurses. By 6:00 P.M. (about 8 hours after collapse) J.D. was able to consume fluids and easily digestible foods. The only signs or symptoms noted the following day were extreme weakness and a general malaise. J.D. was discharged approximately 24 hours after arrival at the hospital with instructions from the physician to refrain from competition for the next week or two until examined by his private doctor. No serum enzymes (AST, ALT, LDH) were recorded initially or thereafter.

The immediate diagnosis upon arrival of the attending emergency room physician was "dehydration and hyperthermia," and then after initial treatment was complete (after about 2 hours) the diagnosis was recorded as "healthy white male with heatstroke." Very interestingly, the incorrect final diagnosis at discharge was "heat exhaustion." It seems quite likely, given the elevated core temperature, the extended loss of consciousness, and the other circumstances noted here, that this was a case of EHS.

This case demonstrates that prompt and appropriate medical attention minimizes the complications associated with EHS, but it also shows that proper medical treatment should extend beyond 24 hours. We recommend that the medical staff monitor other indices of organ function (serum enzymes, including AST, ALT, and LDH, and white blood cells) upon arrival and throughout treatment, especially the morning after the condition presents itself, so that a potentially worsening situation can be identified. Additionally, we recommend a more specific plan upon discharge. The doctor prescribed "no competition." We suggest a slightly more conservative approach of refraining from exercise for 7 days after release from the hospital.

This case is an excellent example of a realistic sporting scenario, in which a driven, healthy, acclimated, fit athlete succumbed to the extreme elements and the competitive spirit that consumes many athletes, soldiers, and laborers. This case illustrates the need for proper event scheduling and management. This race should have been scheduled in the very early morning or in the evening, to avoid the radiative heat load emitted directly from the sun and reflected from the surface of the running track. Additionally, water or a fluid–electrolyte replacement solution should have been provided during the race. The athlete should have more fully hydrated himself on the previous day, a day that included a 10-hour bus trip, a short run, and a 3-hour opening ceremony celebration.

SUMMARY

When the heat dissipation of an exercising person is overwhelmed by harsh environmental conditions and metabolic heat, the consequence may be EHS. This condition often has been attributed to a dysfunctional thermoregulatory system, but it is more appropriate to recognize that EHS occurs when heat accumulation exceeds heat dissipation for an extended period of time. EHS is multifactorial; some cases stem from intense exercise in extreme conditions, while other cases occur when a dehydrated individual tries to sustain moderate to intense exercise, and the demand for cardiovascular maintenance of blood pressure compromises skin blood flow. In still other cases, existing febrile diseases increase the likelihood of EHS, as do low fitness levels and lack of heat acclimation. Each case of EHS seems to have its own unique combination of causative factors. However, the predisposing factors are recognizable, modifiable, and preventable. EHS is a medical emergency that in most cases can be avoided and that, when it does occur, should be properly treated with optimal medical care.

Chapter

Heat Exhaustion, Exercise-Associated Collapse, and Heat Syncope

Lawrence E. Armstrong, PhD, FACSM
Jeffrey M. Anderson, MD

The three illnesses discussed in this chapter share several common characteristics; for example, they all may involve collapse and they all involve the cardiovascular system. Further, substantial overlap exists in their clinical presentation, which may result in misdiagnosis. Therefore, the dual purposes of this chapter are to describe these disorders and to provide keys to recognition and diagnosis. Exertional heat exhaustion and heat syncope occur in hot environments, whereas exercise-associated collapse may occur in any environment.

HEAT EXHAUSTION

The earliest accounts of heat exhaustion, formerly known as heat prostration, were published between 1938 and 1947, and chronicled the experiences of laborers and military personnel. These medical reports (1-4) appeared because numerous military heat casualties had been treated in North African deserts (5) and South Pacific jungles (6, 7) during World War II. Other reports of heat exhaustion originated in the American desert (8), Pakistan (9), and Iraq (10, 11).

These clinical reports distinguished heat syncope from heat exhaustion because the latter disorder involves blood plasma volume depletion (1, 12). Although pure forms rarely occur, heat exhaustion may be associated primarily with dehydration or primarily with salt deficiency (13, 14). Salt depletion heat exhaustion is attributed to inadequate replacement of sweat and urinary salt loss in a hot environment, leading to dehydration and reduced extracellular blood volume, as is seen in hyponatremia (see chapter 6). Chronic cases ordinarily require 3 to 5 days to develop and involve exhaustion. Water depletion heat exhaustion is solely caused by an insufficient fluid intake in a hot environment. Voluntary dehydration, first recognized in the 1940s, is involved, as exercising humans rarely replace as much water as they lose in sweat (15). Anhidrotic heat exhaustion, a chronic loss of sweat gland function, does not involve a salt or water deficiency and therefore will be described in chapter 7.

Heat Exhaustion Terminology

J.S. Weiner and G.O. Horne, two physicians with considerable experience in the recognition of environmental illnesses in hot climates, prepared a classic document in 1958 for the Climatic Physiology Committee of Great Britain's Medical Research Council. Several of the world's leading authorities on heat illnesses oversaw the publication of this document, including W.S.S. Ladell, C.H. Wyndham, and O.G. Edholm. Fourteen distinct heat-related disorders were identified. This "classification of heat illness" (1) arose because confusion existed in the identification and nomenclature of such disorders. At that time, vague and unqualified terms were being applied indiscriminately to a wide range of syndromes encountered in hot environments.

Chapter 2 of this text provides the present nomenclature used to describe heat illnesses, as codified in 1998 by the World Health Organization's International Classification of Diseases (ICD) (16). Numerous publications have utilized the ICD taxonomy, which is very similar to Weiner and Horne's 1958 classification, to distinguish heat exhaustion from heatstroke (17-21) in groups of Marine recruits, laborers, and religious pilgrims.

A confounding factor in this nomenclature debate involves a few clinical publications that ignored the existing ICD categories of heat illnesses and defined heat exhaustion independently. Their definition described the signs and symptoms of ill competitors in specific events. Three of these reports, involving one Australian road race, used the term *heat exhaustion* in reference to athletes who collapsed and concurrently exhibited a rectal temperature greater than 38°C (100.4°F) on admission (22-24). Clearly, such cases could have involved severe hyperthermia, but no distinction between heat exhaustion and heatstroke was acknowledged. One of these reports even claimed that heat exhaustion is potentially fatal (22), a view we strongly disagree with. Equally unconventional definitions of heat exhaustion have involved mountain hikers (25) and tractor drivers (26). Such publications, which ignore ICD categories and apply nomenclature inaccurately, obscure the distinctions that exist among the heat illnesses (see table 4.1) and impede rapid diagnosis and treatment.

Exertional heat exhaustion is defined as an inability to continue exercise in the heat because of cardiovascular insufficiency; it is the most common heat-related disorder in military, athletic, and civilian settings (33-36). Exertional heat exhaustion is an inclusive term that includes both salt depletion heat exhaustion and water depletion heat exhaustion. Confusion sometimes arises when physicians attempt to distinguish exertional heat exhaustion from exertional heatstroke. In comparison to exertional heatstroke, exertional heat exhaustion (mild or moderate dehydration) may be distinguished by three characteristics: a modestly elevated rectal temperature (but below 40.5°C, 104.9°F); absence of central nervous system (CNS) pathology such as reduced mental acuity, confusion, or coma; and normal serum enzyme levels (e.g., alanine aminotransferase, ALT; aspartate aminotransferase, AST; lactate dehydrogenase, LDH; creatine phosphokinase, CPK), which invariably are elevated in cases of heatstroke involving tissue damage or cell death. When a patient exhibits symptoms of both exertional heat exhaustion and exertional heatstroke, rectal temperature should be measured. Whole-body cooling should be instituted immediately if marked hyperthermia is discovered. If rectal temperature cannot be measured promptly (see chapter 3), casualties should be treated empirically. The serum enzyme levels of patients (i.e., soldiers who were running at the onset of heatstroke) have been reported as follows: ALT, 452 ± 294 IU · L^{-1}; AST, 908 ± 553 IU · L^{-1}; LDH, 575 ± 132 IU · L^{-1}; and CPK, 5664 ± 2133 IU · L^{-1} (37). These blood constituents peaked between 1 and 5 days post-heatstroke. Normal values for these enzymes range from 33 to 213, 92 to 186, 7 to 32 and 2 to 45 IU · L^{-1}, respectively, in healthy individuals. Because serum enzyme elevations are minimal in exertional heat exhaustion, and in most febrile states that accompany mental aberration or coma, their measurement may help a physician assess the degree of tissue injury caused by hyperthermia.

Distinguishing Different Types of Heat Exhaustion

Heat exhaustion typically develops during 1 to 5 days of heat exposure, with ample time for electrolyte and water imbalances to occur in unacclimated individuals who have not fully developed the sodium-conserving adaptations of heat acclimation (13, 38-41). Whether primarily caused by water depletion or salt depletion, exertional heat exhaustion involves reduced extracellular fluid volume and insufficient cardiac output to meet the simultaneous demands of perfusing muscle (for exercise) and skin (for heat dissipation). Exertional heat exhaustion also may involve diminished blood pressure, orthostatic intolerance, dizziness, and syncope (fainting). Because of these characteristics, this condition is recognized by the U.S. Army (42) and the Occupational Safety and Health Administration of the U.S. Government (43) as an inability to continue exercise in the heat because of cardiovascular insufficiency.

Table 4.1

Clinical and Physiologic Distinctions of Exertional Heat Illnesses and Exercise-Associated Collapse

	Heat syncope	Exertional heat exhaustion	Exertional heatstroke	Exercise-associated collapse
Description	Fainting in a hot environment.	Fatigue, weakness, and inability to continue exercise in a hot environment.	Severe hyperthermia.	May result from one of many factors: reduced metabolic energy supply, dehydration, temporary malfunction of thermoregulation, or central nervous system dysfunction.
Etiology	Standing in a hot environment or cessation of exercise.	Fluid–electrolyte imbalance combined with heat exposure.	Hyperthermia disturbs cellular function, membrane integrity, fluid balance, and metabolism.	Various etiologies.
Physiological cause	Postural pooling of blood in leg and pelvic blood vessels because of loss of vascular resistance.	Circulatory dysfunction, dehydration.	Great metabolic heat production; reduced heat dissipation because of environmental heat and humidity and clothing.	Various disorders are involved.

Severity	The mildest form of heat illness; a temporary condition.	Not life-threatening.	A life-threatening medical emergency.	During exercise, this may involve a serious medical condition (e.g., cardiac arrhythmia); after exercise, this typically involves postural hypotension.
Critical treatment factors	Elevation of legs and pelvis.	Restore circulatory function and fluid–electrolyte balance.	Whole-body cooling; monitoring of blood pressure and vital signs.	Treat the disorder, signs, and symptoms.
Recovery	Complete recovery within hours; heat acclimation improves responses to heat exposure.	Recovery occurs within 24-48 hours without sequelae; heat acclimation improves responses to heat exposure.	Varies, depending on the duration and severity of hyperthermia and the extent of the resulting organ damage.	Varies, depending on the disorder involved.
References	12, 27	27, 28, 29	14, 28	30, 31, 32

Water depletion heat exhaustion is caused by water loss that is not replaced adequately (29, 40). In the early stage, this condition is characterized by dry mouth, thirst, scant urine volume, vague discomfort, anorexia, rapid pulse, impatience, weariness, sleepiness, and dizziness. Subsequently, individuals exhibit tingling, **dyspnea, cyanosis,** and physical exhaustion. In cases of severe dehydration, patients display marked impairment of mental capacities, incoordination, inability to stand, and may become restless or hysterical. Whereas water depletion heat exhaustion may occur after only a few hours of exercise, salt depletion heat exhaustion typically requires 3 to 5 days before patients become symptomatic. In contrast, cases of exertional hyponatremia caused by overhydration during a single endurance event (see chapter 6) may occur within a few hours.

Salt depletion heat exhaustion is caused by a small dietary sodium (NaCl) intake, large NaCl loss, or both (29, 40). Three clinical grades have been identified:

- early sodium depletion (whole-body Na^+ loss of $0.5 \; g \cdot kg^{-1}$ body weight), associated with fatigue, negligible urinary Na^+, and reduced plasma volume;
- moderately severe sodium depletion (0.5-$0.75 \; g \cdot kg^{-1}$ loss), involving the onset of pallor, nausea, muscle cramps, hemoconcentration, and possible vomiting; and
- very severe sodium depletion ($>0.75 \; g \cdot kg^{-1}$ loss), associated with the characteristics of early and moderately severe sodium depletion plus collapse, systolic hypotension, and possible shock (29).

These three clinical states are approximately equivalent to losses of 4, 4 to 6, and 6 to 10 liters of isotonic saline, respectively.

Case Report 4.1 describes a 24-year-old soldier who experienced salt depletion heat exhaustion, but with a unique etiology (81). This individual experienced large salt losses because he had inherited the genetic disorder cystic fibrosis, which involves high sweat sodium chloride (NaCl) concentrations. His sodium losses were large, and he experienced vascular collapse in a manner consistent with salt depletion heat exhaustion. Laboratory tests confirmed this diagnosis.

Salt Depletion Heat Exhaustion   Case Report 4.1
Associated With Cystic Fibrosis

J.S., a young infantryman, was known to sweat heavily during summer months. He regularly produced a crusty layer of salt on his clothing when he worked for prolonged periods in the heat because he carried hereditary traits for cystic fibrosis. He was transferred to Saudi Arabia and underwent daily heat acclimation. After 8 weeks, J.S. ran a 4.8-kilometer (3-mile) distance during training and collapsed immediately after the run. His water intake had been greater than that of his fellow soldiers during the event. He was admitted to a hospital with nausea, vomiting, dizziness, muscle cramps, and postural hypotension. His serum sodium concentration was 116 mEq · L^{-1}. Function of his liver, pancreas, and lungs were normal. The diagnosis was salt depletion heat exhaustion (see table 4.2). J.S. was given intravenous normal saline and recovered completely within 48 hours.

Table 4.2

Only a Few Characteristics Distinguish Water Depletion Heat Exhaustion From Salt Depletion Heat Exhaustion

Characteristic	Salt depletion heat exhaustion	Water depletion heat exhaustion
Thirst	Not prominent	Prominent
Urine Na$^+$ and Cl$^-$	Negligible	Normal
Plasma Na$^+$ and Cl$^-$	Below average	Above average
Muscle cramps	In most cases	No
Vomiting	In most cases	No
Orthostatic rise in pulse rate	Yes	Yes
Nausea	Yes	Yes
Anorexia	Yes	Yes
Fatigue or weakness	Yes	Yes
Mental dullness or apathy	Yes	Yes
Loss of skin turgor	Yes	Yes
Dry mucous membranes	Yes	Yes
Tachycardia	Yes	Yes

Data from 29, 44, 45.

Two years later, J.S. was transferred to Cyprus, where daily maximum air temperatures averaged 30 to 34°C (86-93.2°F). Initially, his exercise bouts in the heat were limited to 2.5 kilometers (1.6 miles) per day. On the 10th day of his deployment, he had not exercised but was admitted to the hospital with nausea, muscular cramps, dizziness, and postural hypotension. Blood analysis indicated a serum sodium concentration of 128 mEq · L^{-1}. J.S. responded to intravenous saline and recovered within 24 hours. Two clinical tests indicated that the electrolyte levels in his sweat (81-103 mEq $Na^+ \cdot L^{-1}$ and 102-143 mEq $Cl^- \cdot L^{-1}$) were well above the normal range for acclimated, fit individuals (20-60 mEq $Na^+ \cdot L^{-1}$ and 18-53 mEq $Cl^- \cdot L^{-1}$).

J.S. was subsequently restricted to duty in temperate climates. The authors concluded that all cystic fibrosis patients who visit hot climates or exercise should receive regular salt supplementation. (This case report was revised from reference 81.)

Although the signs and symptoms of water depletion heat exhaustion and salt depletion heat exhaustion are similar (27, 45), certain characteristics, shown in table 4.2, allow these conditions to be distinguished. These include thirst, sodium and chloride concentration in urine and plasma, muscle cramps, and vomiting. However, the similarities of signs and symptoms in the two disorders prompted Dinman and Horvath (46) to report that these two forms of heat exhaustion could not be distinguished by physicians who treat laborers in the aluminum industry. This inability to distinguish the two conditions likely was caused by the inability to measure sodium in an industrial infirmary or because sweat, urine, vomitus, and feces involve mixed water and electrolyte losses. In acute fluid depletion cases, such as in marathon footraces, water depletion heat exhaustion is more likely to occur than is salt depletion heat exhaustion. Water depletion heat exhaustion becomes more frequent in long-duration events such as Ironman triathlons.

Recent Research: Exertional Heat Exhaustion in Nonathletes

In recent years, four significant reports have published new clinical and physiological insights regarding exertional heat exhaustion. These studies were conducted during a religious pilgrimage in the desert of Saudi Arabia (47) and within a complex of underground metalliferous mines, at depths of 1,200 to 1,800 meters, in tropical–arid Australia (48-50). Both studies utilized similar criteria to diagnose exertional heat exhaustion (e.g., fatigue or weakness, headache, dizziness, nausea, vomiting, hyperventilation, little sensory change) and recognized that this disorder results from circulatory insufficiency that may include, but is not limited to, syncope. The Arabian study compared echocardiographic images of the hearts of 26 heat exhaustion patients to 31 control subjects. These images showed, for the first time, that heat exhaustion involves tachycardia and high cardiac output with peripheral vasodilation. These cardiac dynamics strongly support our present understanding of the physiological changes that occur as heat exhaustion develops.

The three Australian studies evaluated 65 to 106 cases of heat exhaustion in underground miners and provided several interesting findings. First, body mass index (weight · height^{-2}) was a significant risk factor for heat exhaustion in individuals whose value exceeded 27 kg · m^{-2} (48). Second, heat exhaustion was related to the air dry bulb temperature and air velocity, as shown in figure 4.1. When air temperature was below 33°C (91.4°F) and when air velocity was

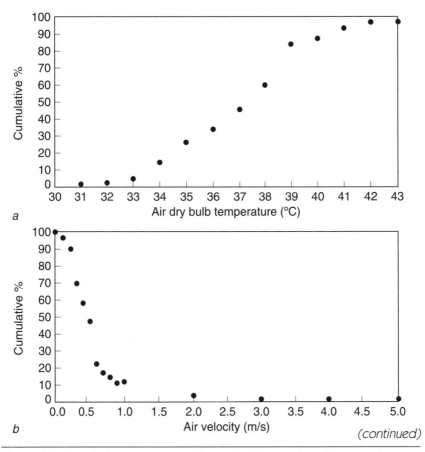

Figure 4.1 Cumulative percentages of exertional heat exhaustion cases in underground metal mines. The incidence of heat exhaustion rose sharply as air dry bulb temperature increased beyond 33°C (91.4°F; panel A, n = 74), air velocity decreased below 2.0 m · s^{-1} (panel B, n = 74), and urine specific gravity increased above 1.010 (panel C, n = 87).

Reprinted, by permission, from A.M. Donoghue, M.J. Sinclair, and G.P. Bates, 2000, "Heat exhaustion in a deep underground metalliferous mine," *Occupational and Environmental Medicine* 57: 165-174.

(continued)

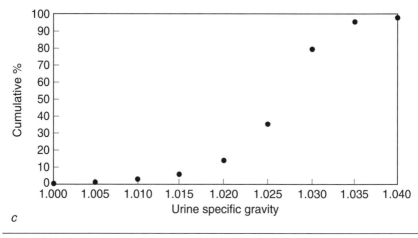

c

Figure 4.1 *(continued)*

elevated (>2.0 m · s⁻¹), heat exhaustion was virtually nonexistent. Third, heat exhaustion was more common when miners were dehydrated (figure 4.1). All heat exhaustion patients had a urine specific gravity ≥1.010. The values of hematocrit, hemoglobin, and serum osmolality also were increased, supporting the concept that heat exhaustion patients were dehydrated. Fourth, intravenous (IV) fluids (1-4 L) were administered to 47% of these patients, usually producing a rapid improvement in symptoms. Patients who preferred oral fluids usually took longer to recover than those who accepted IV fluids (50). Fifth, the incidence of heat exhaustion was greatest during the Australian summer months of January, February, and March (figure 4.2). Sixth, metabolic acidosis occurred in 63% of the cases. This may have resulted from increased anaerobic metabolism, or it may have been compensatory for the hyperventilation and metabolic alkalosis that are often reported as signs of heat exhaustion and heatstroke.

Differential Diagnoses

For exertional heat exhaustion, the diagnosis is one of exclusion (27). More dangerous conditions, particularly heatstroke and hyponatremia, should be adequately ruled out before the diagnosis of heat exhaustion is reached. Especially with increased age of the patient, other basic medical problems, such as heart disease, must be thoroughly considered. If these conditions cannot be ruled out clinically, further investigation is required in an emergency facility.

Although few studies have systematically examined the role of exercise in the appearance of signs and symptoms of exertional heat

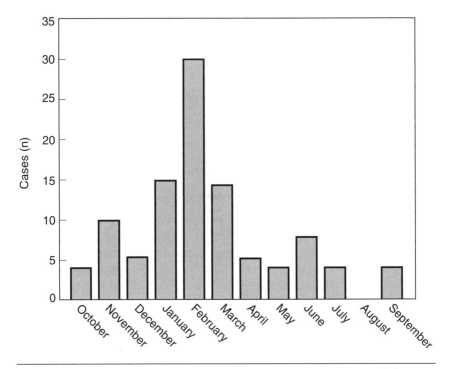

Figure 4.2 Incidence of exertional heat exhaustion by month, among laborers in deep Australian mines.

Reprinted, by permission, from A.M. Donoghue, M.J. Sinclair, and G.P. Bates, 2000, "Heat exhaustion in a deep underground metalliferous mine," *Occupational and Environmental Medicine* 57: 165-174.

exhaustion, Armstrong and colleagues (51) published observations of 14 healthy males who ran intermittently on a treadmill on 8 consecutive days (8.3-9.8 km/day, 63-72 %$\dot{V}O_2$max) in a 41°C (105.8°F) environment. The following signs and symptoms were seen in 18% of the 112 strenuous trials: flushed skin on the head and torso with "heat sensations," 7 occurrences; chills, 7; abdominal cramps, 5; piloerection (i.e., goose flesh), 4; heart rate elevation >160 beats/min after 5 minutes standing upright, 4; "rubbery" legs, 2; vomiting, 2; dizziness, 1; and hyperirritability, 1. This list of signs and symptoms can serve as a guide for diagnosing exertional heat exhaustion.

The assessment of a potentially heat-injured athlete and the severity of illness centers around two pieces of data: the athlete's mental status and rectal temperature (table 4.3). The athlete's mental status should be evaluated first. Frequently, heat exhaustion will cause mild confusion, irritability, incoordination, and fatigue. If an athlete demonstrates more significant alteration in mental status, the medical staff must assume that the athlete has a more serious condition, specifically heatstroke or hyponatremia. Symptomatic hyponatremia arises from overhydration (see chapter 6) and is most often observed in slow endurance athletes who participate in very long duration events. A portable electrolyte analyzer has been used successfully at marathons, triathlons, ultraendurance events, and in wilderness situations, where copious fluid replacement can lead to hyponatremia (52).

Table 4.3

A Comparison of Three Clinical Signs, As They Relate to Heat Syncope, Exertional Heat Exhaustion, Exertional Heatstroke, Hyponatremia, and Exercise-Associated Collapse

Disorder	Syncope	Mental status changes	Core temperature elevation
Heat syncope	Yes	No	Unlikely
Exertional heat exhaustion	Possible	Mild (listless, mild agitation)	Mild (39.5-40.5°C, 103.1-104.9°F)
Exertional heatstroke	Possible	Marked (disoriented, unresponsive)	Marked (>39-40°C, >102.2-104°F)
Hyponatremia	Possible	Marked (disoriented, unresponsive)	Unlikely
Exercise-associated collapse[a]	—	—	—

[a]Exercise-associated collapse (EAC) is a descriptive observation, rather than a pathophysiologic diagnosis. It may result from numerous medical conditions, including hypoglycemia, gastroenteritis, and hyponatremia (32). EAC does not include heart attack, chest pain, heatstroke, insulin shock, or other readily identifiable medical syndromes (30, 54).

After assessment of mental status, the rectal temperature must be obtained. This assessment is often avoided because of lack of privacy or lack of comfort on the part of the examiner or the athlete. Unfortunately, this can be a dangerous oversight. Athletes with exertional heat exhaustion rarely exhibit temperatures above 40°C. If the rectal temperature is higher than this, the medical staff must seriously consider a diagnosis of heatstroke.

Exertional Heat Exhaustion	Case Report 4.2

B.F. was a 43-year-old experienced runner who crossed the finish line after completing a 10-kilometer (6.2-mile) foot race on a warm day. After he walked for a bit, he became light-headed and very weak. He did not collapse but became very nauseated and began to vomit. When emergency responders from the medical team reached him, he was conscious and responsive but was unable to walk without assistance. At the medical area, B.F. was weak and listless but responsive. He recounted no previous medical problems, specifically denying any history of cardiac disease or diabetes. He was a nonsmoker and had no significant family history of cardiac disease. B.F. recounted similar problems after other races. He denied the presence of chest pain or discomfort. He was moderately dyspneic. On examination, B.F. had normal peripheral pulses and blood pressure. Rectal temperature was 39.8°C (103.6°F), pulse was 124 beats/min, respiratory rate was 22 breaths/min, blood pressure was 134/76 mmHg, and he was somewhat pale. The cardiopulmonary examination was unremarkable. A mental status examination showed that B.F. was oriented to person, place, and time and was appropriately responsive.

B.F. was laid supine in the shade of the medical tent, and cold towels were placed around his neck, in his axillae, and in the groin. When his nausea subsided, B.F. was given cool sodium-containing liquids to drink, which he tolerated. A follow-up evaluation revealed that B.F.'s rectal temperature had fallen to 38°C (100.2°F), his pulse to 88 beats/min, and his respiratory rate to 16 breaths/min. His agitation was resolved, and his skin color and turgor had improved substantially.

Because his clinical condition continued to improve, B.F. was discharged in the company of his wife. Upon discharge from the medical tent, he was given the diagnosis of exertional heat exhaustion. His discharge orders were to remain in a cool, shaded area for the remainder of the day, continue his oral fluid intake, salt his food liberally for the next 48 hours, and follow up with his private physician. His physician monitored plasma and urinary sodium levels and recommended that he abstain from strenuous activity for 24 to 48 hours.

Other physical findings associated with exertional heat exhaustion include the preservation of the sweating response but poor **skin turgor** (i.e., the resilience of the skin caused by **capillary filling**). If significant dehydration is present, the athlete's weight will have decreased. The medical staff should ask the athlete what his or her recent healthy weight was; any difference from that weight is likely caused by fluid loss, and the degree of dehydration can be roughly assessed. Along with diminished weight, tachycardia and increased orthostasis are associated with dehydration.

Again, the key physical findings are the patient's mental status and rectal temperature (table 4.3). Regardless of other physical findings, an athlete with a rectal temperature greater than 40°C (104°F) or with altered mental status must be considered a heatstroke victim. When exertional heat exhaustion and heatstroke cannot be distinguished, therapy for heatstroke should be initiated immediately and further laboratory evaluation should be undertaken (27, 53). Laboratory measurements of ALT, AST, LDH, and CPK provide information regarding the degree of tissue disruption. Transaminase elevations are rarely encountered in exertional heat exhaustion; thus, their elevation should trigger concern about the presence of exertional heatstroke.

Heat Exhaustion Treatment

Fortunately, the important work of managing exertional heat exhaustion lies in the accurate diagnosis of the condition. Unlike an athlete with heatstroke, an athlete with heat exhaustion is not in need of rapid cooling. Cooling measures can be applied for comfort and to encourage peripheral vasoconstriction, but no athlete has ever deteriorated from exertional heat exhaustion to exertional heatstroke in a supervised setting (58). The athlete should lie in a cool, shaded location with the feet and pelvis elevated, and restrictive clothing should be removed (59). Cooling measures can include a cool mist spray in combination with a fan to assist in evaporative heat loss. Cool, moist towels may be applied to the head, neck, axillae, and groin. Although full body immersion is the most effective manner of lowering core body temperature (28, 60-63), it typically is not necessary if heatstroke has been ruled out.

The following history should be obtained from each patient: amount of fluid consumed during the race, amount of urine excreted, occurrence of vomiting or diarrhea, medications taken before the event, recent illnesses, amount of carbohydrate ingested before and during the event, and a description of prerace training and heat-acclimation programs (31). Further, fluid losses can be estimated as 1.5 liters (1.6 qt) of water and 2 grams of NaCl per hour of continuous, moderate-to-heavy exercise. This means that exercise lasting 4 hours, in harsh environmental conditions, would result in the loss of 6 liters (6.4 qt) of water and 8 grams of NaCl.

An inexpensive, dilute electrolyte solution (e.g., 0.1% NaCl solution) can be prepared by dissolving two 12-grain salt tablets in one liter (1.1 qt) of water. Palatability is improved if a powdered beverage mix is added to this solution. In addition, commercial fluid–electrolyte replacement beverages provide sodium plus carbohydrates. Oral rehydration is preferred when the patient is conscious unless dehydration is marked, vomiting or diarrhea has occurred, or the athlete is not able to tolerate oral fluids because of gastrointestinal distress (28, 30). In these situations, IV hydration with 500 to 1,000 milliliters of 5% dextrose in 0.5N saline has been used (30). However, 1.0N saline or 5% dextrose saline (with a concentration identical to blood) will replace sodium, and reflect vomiting losses, more appropriately. Before intravenously hydrating a collapsed athlete, one should ensure the absence of clinical signs of symptomatic hyponatremia (54, 64, 65). If the question of hyponatremia is present, a handheld, battery-operated chemistry analyzer can be used to detect hyponatremia (52).

Although **orthostatic hypotension** with no other symptoms may be treated with rest and by elevating the feet (66), exertional heat exhaustion cases involving large fluid losses should be treated with the judicious use of IV fluids (67, 68). This is especially true in cases of severe heat exhaustion, in which serum sodium concentration is greater than 130 mEq $\cdot$ L^{-1} (31, 79). For example, the 1985 Boston Marathon was run on a spring day that was warm (24°C, 75.2°F) in comparison to the environmental conditions that most entrants had trained in (12.5°C, 54.5°F, average daily maximum temperature in Boston during the month of April). Physicians in the finish-line medical tent administered 1 to 4 liters of IV fluid to 158 runners who had been diagnosed with exertional heat exhaustion. Of these, an estimated 90% recovered and walked out of the medical tent within 15 to 20 minutes (68). Retrospectively, it is impossible to determine the runners' levels of dehydration, the role that postexercise orthostatic hypotension played (66, 80), or if these runners would have recovered spontaneously without IV fluids (88).

The following IV solutions have been used to treat heat exhaustion: 5% dextrose (D5W), 0.9% sodium chloride (NaCl) normal saline (NS), 0.45% NaCl (½NS), or 5% dextrose in 0.45% NaCl (D5½NS). Because there is no evidence that one of these is superior (27, 67), the selection of an appropriate IV solution can be left to the judgment of the on-site clinician. In selecting an IV solution, care must be taken to correct serum sodium at a slow rate.

A recent series of investigations, conducted at the University of Connecticut's Human Performance Laboratory, focused on the physiological differences and similarities of various fluid-replacement techniques in healthy, dehydrated test subjects. After dehydration to –4% of body weight, fit males rehydrated with either IV, oral, or no fluid (control) and then cycled to exhaustion in a 37°C (99°F) environment. These studies, published by various authors of the present book (Armstrong, Casa, Castellani, Kenefick, and Maresh), resulted in the following insights. First, the IV versus oral fluids resulted in no differences in body temperature (rectal and skin) or in adrenocorticotrophic hormone (ACTH) and cortisol blood concentrations (69). Second, the exercise time to exhaustion was significantly longer during oral and IV tests than during control tests (i.e., no fluid intake) (70). Third, loss of plasma volume was smaller during oral and IV (versus no fluid) experiments (71). Fourth, oral fluids resulted in lower psychological ratings of perceived exertion, thermal sensation, and thirst compared to ratings with IV fluids (72). Fifth, few differences were observed in the measurements of oral and IV experiments (69, 73, 74). These findings indicate that oral fluid replacement can be utilized in cases of mild dehydration (up to –4% of body weight) without compromising health or recovery. However, IV fluids are appropriate in the treatment of moderately or severely dehydrated individuals (those who have lost more than 4% of body weight). In fact, some authorities have reported that up to 4 liters of IV fluid were required in the treatment of athletes, military personnel, and miners (27, 50, 67, 68).

The volume of required IV fluid can be estimated by measuring serum Na^+ (29, 75). Figure 4.3 presents a sample calculation for a 70-kilogram patient with water depletion heat exhaustion. The calculation assumes a beginning serum Na^+ of 140 mEq $\cdot$ L^{-1} and that total body water (TBW) is equivalent to 60% of body weight. In this case, the patient's dehydration (serum Na^+ = 155 mEq $\cdot$ L^{-1} on admission) indicated that a 4.1-liter deficit had to be replaced by either IV or oral fluids. This approximation underestimates the TBW deficit to the extent that serum NaCl was lost along with water in sweat and urine. Before administering IV fluids, the possibility of preexisting water overload and hyponatremia must be considered (see chapter 6).

Abnormal intestinal fluid absorption also should be considered, according to a study of U.S. Army Reservists undertaking a 2-week summer field training exercise in Texas (19). Four soldiers developed heat exhaustion and voluntarily allowed measurements of body fluids. Deuterium oxide (D_2O) was consumed orally, and blood samples were drawn from these four patients plus two healthy control subjects, at a field medical facility. Normally, TBW in males represents about 60% of body mass. This was similar to the values for the two control subjects (TBW was 58% and 57% of body mass). However, after 4 hours of equilibration, the stable isotope measure-

- Measured serum Na^+ upon examination = 155 mEq $\cdot$ L^{-1}
- Body weight = 70.0 kg
- Assumption: Original serum Na^+ was 140 mEq $\cdot$ L^{-1}
- Assumption: Original total body water (TBW) = 60% of healthy body weight
- $\dfrac{TBW \times original\ serum\ Na^+}{Serum\ Na^+\ upon\ examination}$ = Prevailing volume of TBW

$$\frac{42.0 \times 140}{155} = 37.9\ L\ of\ TBW$$

- Beginning TBW – Prevailing TBW = Water deficit

$$42.0\ L - 37.9\ L = 4.1\ L\ deficit$$

Figure 4.3 Calculation of total body water deficit using the measured serum Na^+ value.

ments of the four heat exhaustion patients indicated that body water represented 70, 83, 86, and 93% of body mass. Because such high percentages of body water are physically impossible, the investigators considered explanations for the differences between patients and control subjects. The most logical mechanism involved disrupted absorption of D_2O from the small intestine into the blood. Because only a portion of ingested D_2O entered the extracellular and intracellular fluid compartments, the TBW pool appeared to be larger than it actually was, and TBW was overestimated. The possibility of impaired intestinal absorption in cases of heat exhaustion has been described elsewhere (27, 39), and concurrent illnesses have been recognized for decades as predisposing factors to heat illness (37, 62). In such cases, oral fluids may not be fully absorbed from the gut, and IV fluids may provide optimal fluid therapy.

Once a patient's signs and symptoms have shown a consistent trend toward resolution, the patient is clinically stable, and he or she feels able to leave, the patient can be discharged in the company of a friend or relative. Prognosis is best when mental acuity and serum enzymes are not altered. The individual should be instructed to follow up with his or her physician (30). Immediate return to exercise or labor is not advisable except in the mildest cases. Affected individuals should allow 24 to 48 hours for recovery. Serious complications are very rare, except when prolonged hyperthermia (>39-40°C, >102.2-104°F) is involved (76). If a patient experiences additional episodes of heat exhaustion, a careful review of nutritional and personal habits should be undertaken and corrective action instituted.

Exertional Heat Exhaustion in Athletes

The preceding text describes different forms of exertional heat exhaustion (e.g., water depletion and salt depletion heat exhaustion) that require hours and days to develop and are caused by fluid–electrolyte deficits. In comparison, athletes competing in single events or training sessions may experience collapse within minutes to hours; this represents a relatively acute physiological disturbance when compared to religious pilgrims who walk in the desert for days, soldiers who live and work in hot environments, or laborers who work five 8-hour shifts per week in a hot–humid industrial setting.

In athletes who exercise for prolonged periods (>4 hours) (66, 77, 78), some clinicians are reluctant to diagnose heat exhaustion, and a few authorities claim that it is not a diagnosable medical condition (58, 66, 78-80), despite the publication of numerous research studies (47-50) and case reports (19, 39, 81, 82) of heat exhaustion during the past 15 years. We believe that this reluctance arises from (a) lack of knowledge regarding the nomenclature contained in the ICD categories; (b) the absence of heat exhaustion in specific populations or situations (e.g., highly trained athletes who compete in ultraendurance events or in contests conducted in mild or cool environments); or (c) lack of knowledge regarding heat exhaustion articles in the clinical and scientific literature.

For example, clinicians and physiologists who believe that heat exhaustion is not a distinct medical condition state that heat exhaustion patients are not necessarily hyperthermic; they believe that the disorder is misdiagnosed (58, 66, 78-80) and actually is orthostatic hypotension. However, the word *heat* was not originally applied to heat exhaustion because patients exhibited an increased core body temperature; rather, it referred to ambient conditions. This disorder most commonly was observed in hot environments. Similarly, heat cramps and heat syncope do not describe a patient's internal hyperthermia, the term *heat* was applied to these illnesses because they commonly occurred in tropical or desert climates (1). Further, some individuals believe that heat exhaustion should be reclassified as a form of exercise-associated collapse. They believe that this allows more efficient diagnosis and treatment in mass participation events (30, 38, 66, 77, 78, 80), even though other sources have encountered exertional heat exhaustion (i.e., as defined in the ICD) during distance running events (83).

Cessation of exercise in the heat may involve several factors, according to observations of 47 healthy heat-acclimated volunteers during 133 exercise bouts (84). The reasons that subjects terminated exercise, without collapse, included dizziness or ataxia (42% of all cases), physical exhaustion (25%), nausea or headache (17%), shortness of breath (12%), and muscle cramps (4%). Those who were classified as heat intolerant were more likely to develop respiratory distress, which is similar to the hyperventilation observed in many cases of heat exhaustion. Heat-tolerant individuals were more likely to continue until cardiovascular insufficiency or illness symptoms limited further effort.

Although anecdotal reports certainly do not prove the etiology of exertional heat exhaustion, they can be illustrative. Such is the case with the experience of Todd Williams, a seasoned distance runner who competed in the 5,000-meter run at the U.S. Olympic Trials in Indianapolis, Indiana, in June 1997:

> I started feeling a little bit of chill and my legs went numb. Then I just went out [fainted]. That's the hardest I've ever run, and the wheels came off.
>
> We were running over our heads for these conditions, but that's what you have to do to improve. I wouldn't want to recommend to everyone that they pass out, but sometimes you have to push the envelope.
>
> Reprinted courtesy of Boston Globe (85).

The meteorological conditions at the time of this event were 35°C (95°F) with 43% relative humidity. Williams and teammate Bob Kennedy, who won the race in 13:31, were attempting to finish between 13:10 and 13:15. This was equivalent to running three consecutive miles in 4:14 each, without stopping.

Williams's comment about "running over our heads" is noteworthy in that exertional heat exhaustion involves the inability of the cardiovascular system to sustain strenuous exercise. When challenged with multiple stresses (such as heat, humidity, dehydration, and high-intensity exercise), the CNS initiates responses in the heart and blood vessels that bolster cardiac output and blood pressure. However, when the intensity and duration of exercise are very great, or when dehydration is great, cardiovascular compensation (i.e., increased heart rate) cannot meet the multiple demands of supplying blood to muscle for exercise, to skin for cooling, and to the brain. Heat exhaustion, and sometimes collapse, eventually occurs unless exercise is voluntarily terminated before collapse (86).

U.S. middle-distance runner Suzy Favor Hamilton, who had run the year's fastest 1,500-meter time in the world in 2000, collapsed during the Summer Olympic Games in Sydney, Australia (87). With about 170 meters remaining in the final event, she collapsed without warning in the midst of runners who were sprinting to the finish line. Her fall was not injury related, she made no attempt to cushion the fall, and she landed forcefully on her left shoulder. To observers, it appeared that she was unconscious at the time of collapse. Both Williams and Favor Hamilton had run similar or faster races previously. But the likely difference from the previous races they had run was the hot ambient temperature of the current race, which placed excessive stress on the cardiovascular system when combined with very high–intensity exercise. In a cool environment, neither of these elite distance runners likely would have collapsed.

Undoubtedly, some authorities would label the experiences of Williams and Favor Hamilton as exercise-associated collapse (30, 54). However, exercise-associated collapse is not a single medical

Exertional heat exhaustion can occur quickly in athletes, sometimes within minutes of the start of competition.

condition but rather a descriptive term that includes many distinct disorders. Does exertional heat exhaustion occur in athletes? Using the time-honored definition—that is, exertional heat exhaustion is the inability to continue exercise in a hot environment because of cardiovascular insufficiency—the answer is yes. As a recognized condition in the ICD (16), heat exhaustion provides a direct diagnosis with treatment options (see chapter 2), whether or not a physician considers exercise-associated collapse first.

Consequences of Fluid–Electrolyte Losses

Some record-setting ultraendurance runners choose to consume very small fluid volumes (e.g., <200 ml) during prolonged competition. Such drinking behavior encourages the belief that dehydration at levels routinely experienced by endurance athletes (–1 to –4% of body weight) does not affect exercise performance negatively. This belief ignores the inability to continue exercise that is observed in exertional heat exhaustion and dismisses dehydration as a contributing etiological factor in this fluid–electrolyte disorder (79, 80). Numerous studies have demonstrated that progressive body water loss (to –5 % of body weight) increases physiological strain in the form of increased heart rate and core body temperature and decreased sweat rate, stroke volume, and cardiac output (88-91). Decrements in outdoor running (92) and cycling performance (70, 93) also have been observed, the former in a mild environment with body water losses equivalent to only 1.6 to 2.1% of body weight. Further, hyperthermia (39.3°C, 102.8°F core body temperature) combined with dehydration results in decreased blood pressure and cardiac output during strenuous exercise (90). Knowing these effects helps us understand the importance of fluid replacement in maintaining cardiac function and leads us to wonder how much faster record-setting ultraendurance athletes would run if they consumed adequate fluid.

In a classic treatise regarding the factors that limit work in a hot environment, Brown (94) described his observations of 22 men who marched 33.8 kilometers (21 miles) through the California desert in 34 to 35°C (94-95°F) heat. This was followed by laboratory tests (cycling exercise, 38.9°C, 102°F) that produced body weight losses of 2.5 to 11%. Brown concluded that exhaustion rarely occurred unless subjects were dehydrated. He also noted that moderate to strenuous exercise by itself generally was not sufficient to initiate collapse. However, once dehydrated, participants could not recover full exercise performance by simply resting or cooling down in a mild environment (95).

To our knowledge, only one published investigation has tracked daily fluid–electrolyte balance before and during the development of exertional heat exhaustion (39). This study involved a large (110.5-kg, 244-lb; 180-cm, 70.9-in.; 32-year-old), healthy soldier (G.A.) during an 8-day heat-acclimation protocol in which he ran intermittently on a treadmill at a self-selected pace for 100 minutes each day in a 41°C (106°F) environment. During the first 7 days of this protocol, G.A. had no clinical signs or symptoms of heat illness and completed the entire 100 minutes of exercise. On day 8, when he should have completed exercise without incident, G.A. instead experienced symptoms of classic salt depletion heat exhaustion: vomiting, fatigue, muscular weakness, and abdominal cramps, without syncope (27, 36, 45). Seven other control subjects completed this 8-day heat-acclimation protocol without incident. On day 8, G.A. also (a) reached the investigator safety limit for rectal temperature (39.5°C, 103.2°F) for the first time, (b) voluntarily ran 2.2 kilometers (1.4 miles) less than the previous shortest day (day 3), and (c) exhibited large elevations of serum cortisol and beta-endorphin concentrations, which indicated a magnified stress response. Table 4.4 presents selected variables that were measured during this investigation. The values for body weight, heart rate, rectal temperature, and plasma volume present a unified concept of the changes that occurred before exertional heat exhaustion. From day 5 to day 8, G.A. lost 5.44 kilograms (12 lb) of body weight. He also experienced an 8-day sodium deficit of 166 mEq, reflected in the elevated plasma aldosterone levels on days 4 and 8, while eating a typical diet of his choice in his residence. The authors of this report concluded that dehydration, caused by salt (i.e., reduced extracellular fluid volume) and water depletion, contributed to the onset of exertional heat exhaustion in this individual.

Circulatory Failure During Intense Exercise in Heat

During exercise in a hot environment, cardiac output increases as much as four times the resting rate (96). As body temperature rises, the brain interprets sensory input from peripheral and central thermal receptors and subsequently causes dilation of cutaneous arterioles and capillaries and increased blood flow to the skin for heat dissipation. As much as 20% of cardiac output may be directed to the peripheral circulation (97). In addition, dehydration and fluid shifts further reduce effective central blood volume, and compensation must occur to maintain cardiac output. After maximal stroke volume is reached, this compensation takes the form of increased heart rate (up to 180 beats/min in hyperthermic individuals) with no change in cardiac output, until exhaustion occurs. In fact, if it were not for intense vasoconstriction of the **splanchnic circulation** (i.e., liver, gastrointestinal organs), inactive muscle, and the kidneys

Table 4.4

Physiological Responses of a Salt Depletion Heat Exhaustion Patient During Daily Heat-Acclimation Trials

Measurement (unit)	Days of laboratory heat acclimation							
	1	2	3	4	5	6	7	8
Beginning body weight (kg)	110.47	109.68	109.58	109.90	112.29	109.40	107.89	106.85
Beginning urine specific gravity	1.024	1.021	1.023	1.024	1.024	1.026	1.026	1.022
Voluntary distance run, at a self-selected pace (km)	8.6	8.1	7.6	8.0	8.4	8.8	7.7	5.4
Heart rate change (beats/min)	79	81	75	65	84	88	77	96
Rectal temperature change (°C)	1.8	2.3	1.8	1.8	1.7	1.7	2.2	2.5
Plasma volume change (%)	−10.7							−16.2
Final plasma sodium (mEq · L^{-1})	139			142				141
Final plasma potassium (mEq · L^{-1})	4.6			4.7				4.7
Final plasma aldosterone (ng · dL^{-1})	42.5			77.2				125.6
Final plasma cortisol (μg · dL^{-1})	11.7			10.0				26.0
Final plasma beta-endorphin (pg · mL^{-1})	40			29				263

Adapted, by permission, from Armstrong et al., 1988, "Heat intolerance, heat exhaustion monitored: A case report," *Aviation, Space and Environmental Medicine* 59: 262-266.

(98), a serious deficit of effective arterial blood volume would occur, and this would lead to profound shock. Obviously, human research studies to evaluate such responses cannot be performed because of ethical considerations.

In describing humans pushing themselves toward maximal performance, Hubbard (99) notes that total peripheral resistance (i.e., vessel tone) fades rapidly and blood pools in the skin and extremities. This pooling of the blood signals that exhaustion and collapse are imminent and that the underlying cardiovascular compensations, supporting muscular contractions and thermoregulation, have failed. Subsequently, a fall in mean arterial pressure indicates cardiac insufficiency. The likely stimulus for this cardiac failure is an excessive reduction of total vascular resistance after the abolishment of compensatory **splanchnic vasoconstriction** (27). Clearly, this involves more than simple postexercise postural hypotension.

Research involving laboratory rats is widely accepted as a valid method of evaluating advanced exercise–heat stress; the regulation of cardiac output in rats is qualitatively and quantitatively similar to that in humans (27, 100). Such animal research has verified the series of events described in the previous paragraph, providing a theoretical mechanism for exertional heat exhaustion in athletes. This mechanism explains not only why heat exhaustion is unlikely in cool environments, but also why collapse in highly trained athletes is sudden and unexpected (i.e., CNS autonomic control of cardiac output changes quickly).

Because the low peripheral vascular resistance and vasodilation cause hypotension, shock is possible. Shock occurs when blood perfusion to the brain and vital organs drastically decreases and is clinically represented by a systolic blood pressure less than 90 mmHg. Peripheral cooling and the resultant vasoconstriction should improve cardiovascular compensation (i.e., increase cardiac output and blood pressure) and will reduce the risk of hyperthermia. By improving blood pressure, cooling also should reduce the risk of overenthusiastic fluid administration. Rectal temperature is likely to be elevated (but not necessarily beyond 40°C, 104°F), because circulatory insufficiency encourages hyperthermia (39).

The signs and symptoms of some patients seem to fall in a "middle ground" between exertional heat exhaustion and heatstroke (27). Mental acuity and rectal temperature are the most important diagnostic criteria in these patients. When a definitive diagnosis is not possible, cooling therapy should be instituted as a precautionary measure against hyperthermia and shock.

Differential Cardiovascular Responses: Water Depletion Versus Salt Depletion

As mentioned, the physiological responses, clinical signs, and patient symptoms of water depletion and salt depletion heat exhaustion differ subtly (table 4.2), depending on the degree of salt or water deficiency. However, water depletion and salt depletion heat exhaustion involve markedly different cardiovascular characteristics; these are relevant to distinguishing exertional heat exhaustion from exercise-associated collapse.

Dehydration caused by water deprivation and a large fluid loss is characterized by thirst and scant urine volume. It is completely relieved by administration of pure water. Although this imbalance results in decreased extracellular fluid volume, plasma volume, and cardiac output, it does not produce peripheral vascular collapse (101, 102).

Dehydration caused by a negative sodium (Na^+) or salt ($NaCl$) balance may be the result of a low dietary intake, large losses of $NaCl$ in sweat, or both. This deficiency results in a loss of extracellular fluid and a reduction of plasma volume, cardiac output, and blood pressure. It is not relieved by ingestion of water, and thirst is not prominent. Severe salt depletion dehydration produces a form of peripheral vascular collapse closely resembling traumatic shock (101, 102). Clearly, this involves a more serious medical condition than simple postexercise postural hypotension. If a laborer or athlete loses a large volume of sweat during several consecutive days of work or exercise, without dietary sodium replacement, significant cardiovascular changes (e.g., loss of peripheral vascular resistance, hypotension) can occur. Thus, salt depletion heat exhaustion (see chapter 2), a fluid–electrolyte disorder that is included in the ICD (16), may be responsible for some forms of collapse in athletes. If the salt deficiency is severe, circulatory shock may occur, without exercise or any element of trauma (102). Case reports 4.1 and 4.2 presented examples of this fluid–electrolyte disorder.

The cardiovascular effects of sodium deficiency are especially noteworthy. Because blood pressure may not be decreased initially, evaluation of orthostatic changes of blood pressure and pulse rate are necessary in patients who are evaluated for volume depletion. Before advanced salt deficiency develops, postural hypotension may exist and the pulse rate may be elevated; this may occur before a decrease in supine blood pressure takes place (45). Dry mucous membranes, decreased skin turgor, and sunken eyes and cheeks suggest significant sodium depletion and dehydration.

EXERCISE-ASSOCIATED COLLAPSE

Because exercise-associated collapse (EAC) may occur in a cool environment and may be caused by numerous medical conditions, triage at the point of collapse should result in a firm diagnosis based on the patient's history and clinical evaluation. EAC may involve hypoglycemia (i.e., symptoms of weakness, headache, hunger), nausea, vomiting, abdominal pain, loss of appetite, or hyponatremia (see chapter 6) (32). By definition, EAC does not include heart attack, chest pain, heatstroke, insulin reaction or shock, or other readily identifiable medical syndromes (30). This condition is characterized by abnormal body temperature (i.e., hypothermia or hyperthermia), altered mental status, muscle spasms, and inability to walk or ingest oral fluids (30).

Several risk factors for EAC have been identified. These include inadequate training, failure to carbohydrate-load, failure to eat breakfast on the morning of an event, prerace illness (especially one involving a fever), and hypothermia in cool, windy, and wet conditions (30). After exercise ceases, EAC often involves orthostatic hypotension or syncope, as described in case report 4.3. In contrast, the signs and symptoms of heat exhaustion may include headache, hyperventilation, flushed skin on the head and torso, chills, nausea, vomiting, abdominal cramps, piloerection (i.e., goose flesh), and persistent postexercise heart rate elevation, with a heart rate of greater than 160 beats/min while in an upright posture (27, 39, 44).

Exercise-Associated Collapse	Case Report 4.3

T.R. was a 20-year-old track-and-field athlete, competing in her conference championships as a heptathlete and 400-meter hurdler. On the second day of the competition, after a personal record in the hurdles, T.R. crossed the finish line, stopped abruptly, and then collapsed. When the sports medicine staff reached her, T.R. was conscious with normal, spontaneous respirations and a strong pulse. The staff assisted her to the medical tent. She had not been drinking excessively during the day. The team's athletic trainer reported that T.R. had a history of similar episodes, without hyperthermia. In a previous evaluation, she had exhibited a normal electrocardiogram, a normal echocardiogram, and a normal maximal exercise treadmill test. She had been under the care of a nutritionist because of inadequate caloric and fluid intake but had not been diagnosed with an eating disorder.

Upon examination, T.R.'s rectal temperature was 37.7°C (100°F), her pulse was 116 beats/min, her respiratory rate was 20 breaths/

min, and her blood pressure while supine was 120/70 mmHg. Examination of her head, eyes, ears, nose, and throat revealed minimally dry mucous membranes. The heart, lung, and abdominal examinations were unremarkable. Radial pulses were strong in both wrists, and dorsalis pedis pulses were strong in both feet. A mental status examination found T.R. to be responsive, though somewhat sluggish. She was oriented to person, place, and time. She followed commands adequately, and she had appropriate recall of recent and remote events, including the results of her competition on that day. The diagnosis was mild normothermic exercise-associated collapse of nonspecific origin, related to intense exercise.

Her treatment included rest in the supine position in the medical tent. T.R.'s feet and pelvis were elevated, and she was allowed to take oral fluids as she tolerated. Her vital signs were rechecked, and both her heart rate and respiratory had slowed appropriately. When she had recovered her strength after about 10 minutes, T.R. was allowed to leave the treatment tent with a teammate.

A common misconception suggests that a syncopal athlete is always dehydrated (78). However, collapsed athletes are neither more dehydrated nor more hyperthermic than their peers who do not collapse (58). Additionally, Roberts, in his review of 12 years of injury in the Twin Cities Marathon (103), noted no predisposition to hyperthermia in athletes suffering from exercise-associated collapse in cool conditions. The signs and symptoms of collapsed athletes include some of those found in exertional heat exhaustion. Further, some runners at the Twin Cities Marathon, despite the cool environmental conditions, have elevated body temperatures and are distinguished only by a rectal temperature measurement.

The physiological responses underlying EAC have been clarified (54, 58, 83, 104). When exercise ceases, as at the end of a long run, the muscle pump ceases abruptly. However, the peripheral vasodilation that occurs with exercise changes slowly. Additionally, the increased parasympathetic tone and diminished sympathetic activity that occurs with physical training (e.g., 8-12 weeks of endurance exercise) may blunt the vasoconstrictor response in athletes and may trigger paradoxical vascular engorgement of the splanchnic vasculature (78). Thus, blood pools in the periphery and does not return to the central circulation, diminishing perfusion to the brain. Ultimately, syncope ensues.

Clinicians frequently assume that a collapsed athlete is dehydrated and is in need of IV fluid replacement. Additionally, the diagnosis of heat exhaustion erroneously may be assumed, when the actual signs and symptoms imply a clinical scenario that is quite different. These misconceptions have led some authorities to imply that hydration status and heat are unrelated to athlete collapse. These authorities have drawn conclusions from endurance events that were conducted in mild or cool conditions (54, 58, 78, 103, 104). In reality, dehydration, elevated ambient temperature, or elevated core temperature need not be present for EAC to occur. In fact, discovery of an elevated core temperature (or other signs and symptoms that are not seen in simple orthostatic hypotension, such as headache, nausea, vomiting, and hyperventilation) may shift the diagnosis to heat exhaustion or heatstroke, depending on the degree of core temperature elevation and the presence or absence of more serious mental status changes. On the other hand, the occurrence of EAC likely is not completely independent of hydration status and ambient temperature. Instead, although neither of these factors is the sole, or even the primary, determinant of EAC, they both play a physiological role in the development of the disorder.

Exercise in high ambient temperatures results in increased fluid loss from sweating, and the resulting hypohydration causes plasma volume to diminish. This plasma volume loss further decreases the central venous pressure, the left ventricular end diastolic volume, and ultimately perfusion pressure to the head. These effects are blunted by the body's physiological response to hypohydration. In the hypohydrated athlete, skin blood flow is restricted and sweating is reduced. These responses counter the effects of hypohydration on the central circulation and are the reasons that heat dissipation is impaired by hypohydration. However, if the fluid or heat stress on the athlete is great enough, these compensatory measures will be insufficient to fully buffer the effect of hypohydration on the central circulation and ultimately on perfusion pressure to the brain. When hypohydration is combined with the rapid discontinuation of the muscle pump at the end of exercise, the possibility of syncope is increased. Through this mechanism, fluid loss and elevated ambient temperatures increase the risk of EAC and heat syncope.

Assessment and Management of Exercise-Associated Collapse

Management of the EAC patient begins by ensuring an open airway, spontaneous breathing, and the integrity of the patient's circulatory capabilities. This corresponds with the basic ABCs (i.e., airway, breathing, circulation) of an emergency medical response. Usually, the EAC patient will not lose consciousness, making this assessment elementary. After assessing the ABCs, the EAC patient should be moved or assisted to a safe area. This condition often occurs at the finish line of a distance running event, and the risk of injury is greater from trauma caused by the crowd or a fall than it is from the condition itself.

Through a medical history inquiry, the medical staff should determine if the athlete drank too much fluid during the event. The presence of muscle twitching, muscle fasciculation, or epileptic seizures suggests hyponatremia (see chapter 6). Progressive clouding of consciousness also suggests hyponatremia, hypoglycemia, hyperthermia, and hypothermia (78, 79).

The EAC patient should be placed in a cool, shaded area and laid supine. The legs and pelvis should be elevated to maximize perfusion to the brain. If medical personnel feel compelled to provide additional care for the patient, EAC patients may be treated with cooling, warming, or rehydration measures. Because the condition is not caused by elevated core temperature or dehydration, these measures are not necessary. As long as the patient's blood pressure, pulse, and rectal temperature are normal, IV fluids are not required. Similarly, ice packs and cool water immersion are not necessary, but they will enhance total peripheral resistance in blood vessels, cardiac output, and central blood volume. Once cerebral perfusion is restored and normal vascular tone has returned, the EAC patient can be allowed to leave the medical treatment area at his or her leisure.

Possible Etiologies of Exercise-Associated Collapse

If syncope occurs after exercise, when environmental conditions are mild and skin vasodilation is normal, the proper diagnosis is EAC, not heat syncope. Symptoms of EAC, according to a checklist developed at the Twin Cities Marathon in Minnesota (30), may include one or more of the following: exhaustion, fatigue, light-headedness, confusion, headache, nausea, stomach cramps, leg cramps, feeling hot, or feeling cold. In EAC cases, one or more of these signs may be observed: abnormal body temperature, muscle spasms, altered mental status, unconsciousness, CNS changes, inability to walk without assistance, vomiting, tachycardia, and diarrhea.

In addition to the EAC paradigm developed by Roberts (30), other possible causes for collapse during exercise include cardiac and noncardiac causes. Cardiac-related causes include heart valve obstruction abnormalities (e.g., arrhythmias) and nerve dysfunction. Noncardiac causes include metabolic (e.g., low blood sugar), neurological (e.g., seizure), and psychiatric causes (105). However, **vasovagal syncope** probably is the most common type of fainting experienced by athletes. The term *vasovagal* refers to the effect of the vagus nerves (i.e., passing from the brain downward through the neck) on veins and the heart. In response to prolonged standing, dehydration, exercise, and an increased blood epinephrine concentration, these nerves initiate a reflex in the brain that slows the heart (via the parasympathetic nervous system) and dilates peripheral blood vessels (via the sympathetic nervous system). As a result, cardiac output and blood pressure decrease in concert with pallor, nausea, blurred vision, and fainting (106). Although this type of collapse is not life-threatening, the falls resulting from such vasovagal events can be dangerous. When athletes repeatedly experience syncope, cardiac complications or disease likely are involved. With proper diagnosis and medications, such individuals may be able to return to training and competition (107).

When evaluating patients who have collapsed, causes of syncope other than EAC also should be considered. Table 4.5 distinguishes these other types of syncope from heat syncope. The differential diagnosis for syncope resides mostly in the realm of neurally mediated, or cardiac, disturbances (55). The most common cause of syncope in the general population is vasovagal, or neurocardiogenic, syncope (55-57). Other possible causes of syncope include structural heart disease (such as coronary artery disease, congenital heart disease, heart failure, or valvular disease), arrhythmia, seizure, or psychiatric disorders

The most elusive and concerning of these diagnoses is cardiac arrhythmia. Because of its potentially fleeting nature and paucity of physical findings, an underlying cardiac arrhythmia cannot entirely be ruled out as the cause of syncope. However, the clinical situation and likelihood ratios can be helpful in appropriately assessing the athlete. In a young athlete with no underlying heart disease who experiences syncope that matches the clinical scenario associated with EAC, the likelihood of underlying cardiac arrhythmia is extremely low. The exclusion of arrhythmia becomes more difficult with advancing age of the patient and the presence of underlying cardiac risk factors. A syncopal athlete with known cardiac disease, or one who experiences syncope during exercise, likely has an underlying cardiac arrhythmia and should be treated accordingly.

HEAT SYNCOPE

Although it may coexist with exertional heat exhaustion, heat syncope is categorized as a distinct syndrome in the ICD (16) because it may occur in unacclimated individuals who have normal fluid–electrolyte status. Heat syncope refers to a fainting episode (i.e., a form of collapse) and usually involves unfit, sedentary, unacclimated individuals who stand erect for a long period of time and wear a uni-

Table 4.5
Clinical Features That Distinguish Various Types of Syncope

Causes of syncope	Differentiating clinical features
Coronary artery disease or myocardial ischemia	Chest pain or pressure; presence of risk factors (male gender, age >35, diabetes, smoking, hypertension, family history); electrocardiogram abnormalities.
Arrhythmia	Recognized underlying heart disease; usually older age; may have prolonged syncope or require resuscitation. Strongly consider this diagnosis if syncope occurs during exercise.
Vasovagal, or neurocardiogenic, syncope	Previous occurrence, often beginning at a young age; related to posture; occurrence after limited food and fluid intake; related to warm or hot ambient temperatures; associated with emotion, urination, cough, or defecation.
Seizure	Witnessed seizure activity; loss of bowel or bladder control; tongue biting; use of medication that lowers seizure threshold; period of confusion after the syncopal episode (postictal state).
Medications	Use of insulin that causes hypoglycemia, central acting agents that lower blood pressure, or medications that prolong the QT interval (electrocardiogram) or cause slow heart rate.
Psychiatric disorders	History of generalized anxiety disorder or panic disorder, major depression, or substance abuse.
Autonomic instability	Presence of an underlying illness that affects the autonomic nervous system (e.g., diabetes is the most common).
Heat syncope	Fainting due to hypotension that results from lack of heat acclimatization, upright posture, and gravitational pooling of blood in the legs, feet, and pelvic girdle.

Data from 55, 56, 57, 107

form or insulated clothing. School band members and military personnel, during their initial exposures to elevated air temperature, comprise the classic patient populations. Whereas the EAC patient stops exercise abruptly and remains upright, the heat syncope patient usually does not exercise. An individual experiencing heat syncope will have no seizure activity or cardiac rhythm abnormalities, and the individual will recover mental status relatively quickly.

Heat syncope occurs during the initial days of heat exposure because the unacclimated cardiovascular system has not adapted to heat stress. Physiologically, the high air temperature and clothing insulation result in maximal skin vasodilation and pooling of blood in the veins of the extremities. This pooling causes a decrease in the volume of blood returning to the heart. Because cardiac filling decreases, the cardiac output declines, resulting in a marked decrease in blood pressure and insufficient oxygen delivery to the brain.

Full heat acclimation requires 8 to 14 days (108) and represents a complex of physiological adaptations that reduces physiological strain. Figure 4.4 illustrates the incidence of heat syncope among 45 subjects, who lived in a hot environment for 9 days, 24 hours each day (109). After the first day, the incidence of heat syncope clearly declined with each additional day of exposure to heat. This decline in syncopal events paralleled the classic circulatory adaptations that occur during the first 3 to 5 days of heat acclimation (108), including plasma volume expansion, decreased heart rate, and improved orthostatic tolerance (110). These results support the concept that heat syncope involves altered cardiovascular control. In subjects who do not collapse, a compensatory increase of both heart rate and cardiac

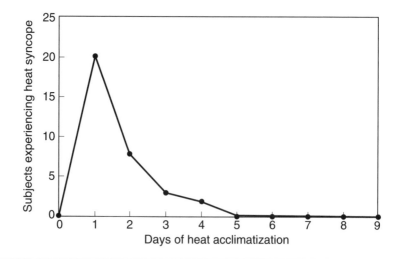

Figure 4.4 Incidence of heat syncope among 45 test subjects living in a hot environment for 24 hours each day and undergoing exercise trials.

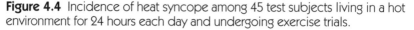

Adapted, by permission, from W.B. Bean and L.W. Eichna, 1943, "Performance in relation to environmental temperature," *Federation Proceedings* 2: 144-158.

output maintains blood pressure at levels adequate to maintain consciousness (4).

Medical Considerations of Heat Syncope

Although it is necessary to distinguish heat syncope from EAC and other serious causes of syncope, the evaluation should begin by seeking an eyewitness account and examining the patient's history of similar episodes. Immediately before fainting, the heat syncope patient may experience sensations of weakness, tunnel vision, vertigo, or nausea. The medical diagnosis of heat syncope is based on a brief fainting episode in the absence of elevated rectal temperature. If rectal temperature is greatly elevated, heatstroke should be suspected (see chapter 3). The pulse and breathing rates are slow and skin pallor is obvious. The patient should be examined for head injury, dehydration, anemia, urinary incontinence, urine specific gravity, and urine sodium concentration.

After heat syncope occurs, a horizontal posture encourages venous blood to return to the heart, causing an increase in cardiac output, blood pressure, and delivery of blood to the brain. Recovery of consciousness seldom takes more than a few minutes. In a hot environment, complete recovery of blood pressure, pulse rate, and the sense of well-being often requires 1 or 2 hours. The heat syncope patient often remains pale, is considerably shaken by this experience, and, after reassurance, usually prefers to rest or sleep. These characteristics demonstrate that heat syncope is a more profound disturbance of circulatory stability than is ordinary postural hypotension (12).

First aid and medical treatment for heat syncope include resting in a cool or shaded environment with feet elevated, avoiding sudden or prolonged standing, and replacing fluid and electrolyte deficits. Movement of cool air across the skin (e.g., fanning) and cold water immersion reflexively influence muscle tone and cutaneous vasoconstriction such that venous return to the heart and blood pressure return to normal (4, 111). Unacclimated individuals should be instructed to grade the duration and intensity of their activities during the next 8 to 14 days (108), so that they do not prematurely attempt strenuous exercise or labor (12). Similarly, daily exertion should be modified, even for acclimated individuals, if there is a sudden increase in environmental temperature (108) or humidity (112).

SUMMARY

According to the clinical literature, exertional heat exhaustion, heat syncope, and EAC appear in different situations. The two former illnesses occur in hot environments whereas the latter occurs during endurance events conducted in mild environments. Because their signs and symptoms overlap considerably, diagnosis is difficult and confusing. However, delimiting diagnoses are not of great importance because these disorders are benign and can be treated similarly. Although definitions vary among clinicians, heat exhaustion involves neither the loss of mental acuity nor hyperthermia exceeding 40.5°C (104.9°F). EAC is characterized by abnormal body temperature, altered mental status, muscle spasms, and inability to walk or ingest oral fluids. Heat syncope involves hypotension and a brief fainting episode, in the absence of elevated rectal temperature. These three illnesses must be distinguished from exertional heatstroke and exertional hyponatremia before treatment is administered.

The authors gratefully acknowledge the thorough review comments of William O. Roberts, MD.

Chapter

Exertional Heat Cramps

Michael F. Bergeron, PhD, FACSM

A Common Problem; A Simple Solution   Case Report 5.1

He was not unlike other young, fit, and talented 17-year-old tennis players. A.H. trained hard, competed regularly with the best, and attained a respectable national ranking. He was accustomed to doing well in the early rounds of most tournaments. Unfortunately, after winning several matches, particularly in events characterized by hot weather and heavy sweat losses, A.H. often faced an unyielding opponent that many of his counterparts seemed to avoid: exertional heat cramps.

These debilitating muscle cramps, which primarily affected A.H.'s legs, occurred despite his efforts to hydrate and eat well between matches. Although A.H. experienced heat cramps more often in hot and humid weather, high heat and humidity were not always prerequisites. A number of trainers and physicians examined A.H., and he received and tried a variety of ineffective "remedies." Notably, A.H.'s father had high blood pressure; consequently, the entire family regularly consumed a diet fairly low in salt. Subsequent testing of A.H. revealed high rates of fluid and sodium loss (via sweating) during competitive singles play in the heat—2.5 liters and 89.8 millimoles (just over 2 g) *per hour,* respectively. With an increase in daily salt intake, along with careful attention to appropriate fluid intake throughout the day, A.H. was able to completely avoid exertional heat cramps during subsequent competition, even during events where the incidence of heat cramps for other athletes was high (1).

Athletes work hard to increase their muscle size, strength, and power, so that voluntary efforts to contract selected muscles effectively contribute to optimal performance. However, an unexpected prolonged series of involuntary muscle contractions (cramps) can readily turn a well-developed, powerful athlete into a feeble victim writhing in pain. Muscle cramps and competitive sports seem to go hand-in-hand. Although muscle cramps have a variety of etiologies, the discussion in this chapter is limited primarily to muscle cramps that occur during or after extensive exercise and sweating. These are the exertional heat cramps.

Athletic performance and risk for heat illness are directly related to hydration status (2-5), and most athletes, coaches, and trainers appreciate the value of drinking fluids to offset sweat losses during competition or training. The importance of adequate hydration during exercise has been further emphasized by several national associations (6-8). Nonetheless, many athletes experience exertional heat cramps. Adequate fluid intake can help, but even when a lot of water or other low-sodium fluids are consumed, heat

© Paul McMahon

Exertional heat cramps can occur when an athlete has not properly acclimatized herself to the environment, not consumed enough water, or has not followed good dietary guidelines, including eating enough salt.

cramps sometimes still occur. Clearly, then, other factors contribute to this illness.

This chapter begins with a discussion of the etiology and related pathophysiology of heat cramps. Common scenarios that seem to prompt a higher incidence of heat cramps and factors that predispose athletes to experiencing heat cramps are then highlighted. This is followed by a look at signs and symptoms (including early warning signs), treatment, and continuation of activity and return to play considerations. However, a primary and practical aspect of this chapter is a comprehensive discussion of prevention of exertional heat cramps. Most, if not all, athletes can avoid exertional heat cramps if they take appropriate and adequate preventive measures before and during competition. Moreover, the information and recommendations presented here are not exclusive to athletes; laborers, military and service personnel (e.g., firefighters), and others who sweat considerably often are impaired by heat cramps as well.

ETIOLOGY AND RELATED PATHOPHYSIOLOGY

Many factors have been proposed as the cause of muscle cramps during exercise. Among the suspected culprits are a variety of mineral deficiencies—calcium, magnesium, and potassium (9-18). A deficiency in one or more of these minerals or a related cellular dysfunction or pathophysiology (19) could certainly lead to muscle cramps or various other neuromotor problems. Accordingly, such mineral supplementation may be warranted in some cases to avert such anomalies. However, when an athlete cramps in the heat, these particular minerals usually do not appear to have been lost in great quantities or lacking in the diet (Bergeron, unpublished findings). Insufficient conditioning and fatigue can also cause muscle cramping in an overworked muscle (20, 21), but such cramps are usually localized and can be resolved through passive stretching, massage, or icing. Not so with heat cramps.

In contrast to these potential causes of exertional heat cramps, extensive and repeated water and sodium losses often have been cited as primary risk factors for the onset and development of heat cramps (1, 9-11, 13, 17, 22-24) and are seemingly the contributing determinants. Without adequate sodium and fluid replacement (25), progressive and significant sodium and body water deficits can result. Consequently, the extracellular fluid compartment of the muscles can become contracted (26). This, in turn, can cause selected motor nerve terminals to become hyperexcitable and to seemingly spontaneously discharge, due to mechanical deformation of these nerve endings and changes in the ionic and neurotransmitter concentrations of the surrounding extracellular spaces (23, 27).

COMMON SCENARIOS

Heat cramps seem to occur most often during or following prolonged exercise when there has been profuse sweating and, consequently, extensive water and electrolyte (primarily sodium and chloride) losses. When it is hot, athletes and coaches usually recognize the occurrence of sweating and ensure that a greater degree of fluid intake and balance are maintained. However, the full extent of the fluid losses is often unappreciated. Most adult athletes can readily lose between 1 and 2.5 liters of sweat during *each hour* of intense training or competition, especially when the heat index is high. Some athletes lose even more. In fact, sweating rates have reportedly exceeded 3.5 liters per hour in hot and humid climates (28). Thus, the potential for a significant body water deficit during a long race, game, match, or training session is evident. In addition, even with a low concentration of sodium in the sweat (see Prevention on page 98), such high sweat rates will usually yield rather impressive sweat sodium losses: losses of 2,000 to 5,000 milligrams of sodium per hour are common among many athletes and recreational exercisers. Without adequate sodium replacement, a sizable deficit in total body exchangeable sodium could readily develop.

Particular athletic events, such as marathons, cycling events, and soccer or football games in warm weather, lend themselves to extensive sweating and mineral loss. However, participation in some sports, such as tennis, where a player may have to compete more than once a day on several successive days, can prompt an even greater risk for incurring heat cramps, because of the extensive repeated sweating and inadequate rehydration efforts or opportunities. In junior tennis, regulations for typical situations state that a player must have at least a 1-hour rest before a subsequent match can begin (29). If the previous match was long and there was considerable sweating, 1 hour is likely not enough time to adequately restore fluid and electrolytes that were lost through sweat. Thus, the player often begins the next match with a clinically significant fluid and sodium deficit. It is during this second match on the same day that certain players are at greater risk and often fall victim to heat cramps, whereas they had no related difficulty during the previous contest. A similar scenario could readily develop, for example, with football practices during "two-a-days" or during a soccer tournament. Of course, even in a single athletic event such as a marathon, if the duration is long enough and sweating is extensive enough, the risk for incurring heat cramps increases as a progressive fluid and sodium deficit develops.

Importantly, it does not have to be hot for athletes to lose extensive fluid and electrolytes from sweating. Heat cramps can also occur during indoor competitions or in cool weather. In addition, heat cramps can develop even when an athlete has been regularly drinking plenty of fluid, especially when the type of fluid is water or some other low-sodium drink and there is no other concomitant intake of salt via food (see Prevention on page 98 for examples of appropriate fluid replacement strategies). In fact, in the presence of a sweat-induced sodium deficit, repeated ingestion of too much low-sodium or sodium-free fluids (e.g., water) will not only fail to prevent or resolve heat cramps, but could lead to hyponatremia (22, 30-33) and other dangerous clinical symptoms (34). A comprehensive discussion of hyponatremia can be found in chapter 6.

SIGNS AND SYMPTOMS

Those who are susceptible to and often afflicted by heat cramps might anticipate a particularly higher risk on a hot and humid day, especially if the duration of activity is expected to be long. However, an athlete's first sensory indication of the onset of heat cramps typically begins with very subtle, almost undetectable, "twitches" or **fasciculations** (23) in certain voluntary muscles. These twitches may first appear in the quadriceps, for example, and then subsequently somewhere else. Unless the ensuing problem is appropriately and immediately addressed (see Treatment on page 96), the athlete will face full-blown heat cramps.

If and when heat cramps actually develop, only a few localized muscle bundles in a given muscle contract at a time. However, adjacent bundles subsequently contract, as other muscle fibers begin to relax. Electromyography indicates periodic high-voltage discharges that spread from one area of the afflicted muscle to another, with regional irregular fascicular twitches occurring between cramps (23). Thus, athletes often report that heat cramps seem to "wander" (35). Despite the fact that the entire muscle mass is not affected all at once, athletes will readily admit (and outwardly display) that the pain of heat cramps can be excruciating and debilitating.

When an afflicted athlete has a blood test to examine electrolyte levels and other diagnostic indicators, a normal or slightly elevated serum sodium concentration often mistakenly suggests that there is no sodium deficit. Conversely, lower serum sodium or even slight hyponatremia may simply be an acute response to rapid or repeated ingestion of low-sodium or sodium-free fluids such as water. Measured plasma or serum electrolyte levels may not necessarily indicate the presence or absence of a sodium deficit (36). This is particularly true during or right after intense exercise. Intense exercise will cause a change in plasma volume (hemoconcentration), and the athlete will secrete sweat that is hypotonic compared to plasma; this combination may help maintain or actually slightly elevate circulating sodium levels. Moreover, once an athlete stops exercising and begins to appropriately rehydrate and replace lost electrolytes, serum sodium is typically normalized very quickly, so that tests done even soon after exercise may not reflect a sodium deficit if it exists.

To obtain a truer perspective on the potential for or degree of an existing total body exchangeable sodium deficit, a careful assessment and comparison of recent dietary sodium intake with sweat electrolyte losses is much more informative. Figure 5.1 illustrates calculations that can be made when laboratory analyses for sweat and urine concentrations of sodium are possible. Dietary analyses can be performed with nutritional software (e.g., Nutritionist Pro™) that provides dietary sodium analysis. The calculations in figure 5.1 show that dietary sodium replacement requires careful planning for an athlete with a high sweat sodium concentration. The values in figure 5.1 represent an actual evaluation of a 17-year-old tennis player. Extreme urine sodium conservation, as determined from a 24-hour urine collection, also would likely be more indicative of a significant sodium deficit than circulating sodium levels would be.

Other clinical signs and symptoms (e.g., headache, fatigue, and slight nausea) are sometimes evident with heat cramps. However, these symptoms may be related more closely to a coincident state of heat exhaustion or hyponatremia than to the heat cramps.

TREATMENT

If an athlete recognizes the symptoms early enough—that is, when the subtle muscle twitching is first noticed—the impending episode of muscle cramps and debilitation can be averted. Consuming a solution of about 0.5 teaspoon of salt (NaCl) dissolved in 16 to 20 ounces of water or sports drink (if the athlete can tolerate this concentration) right away can effectively prevent any further development of twitches or cramping (1). Even if moderate heat cramps are actually present, consuming such a mixture could relieve the cramping. Alternatively, intravenous rehydration may be necessary and more promptly effective (e.g., 0.5-1.0 L normal saline). Salt tablets are also effective in treating (or preventing) heat cramps, as long as they are taken with plenty of fluid, such as with water or a sports drink. Three salt tablets (1 g of NaCl per tablet) will provide the same amount of sodium as 0.5 teaspoons of salt.

Quinine continues to be a customary option in treating heat cramps, despite empirical evidence suggesting varying levels of efficacy, as well as no benefit (37). Quinine reportedly decreases motor end plate excitability, while increasing the muscle membrane refractory period. Can such effects alter or reduce neuromotor activity, such that overall athletic performance is diminished? Perhaps.

Sweat sodium and sodium chloride loss during strenuous exercise in the heat

Given: Duration of exercise = 4 h of tennis match play
   Sweat rate = 2.5 L · h^{-1}
   Sweat sodium concentration = 83 mEq · L^{-1} [a]
   1 mEq Na$^+$ = 23 mg Na$^+$
   1,000 mg Na$^+$ = 1 g Na$^+$
   1 g NaCl (sodium chloride) contains 393 mg Na$^+$

Total sweat loss: (2.5 L · h^{-1}) × (4 h) = 10 L sweat lost

Total mEq sodium in sweat: (10 L) × (83 mEq Na$^+$ · L^{-1}) = 830 mEq Na$^+$

Total mg sodium in sweat: (830 mEq Na$^+$) × (23 mg Na$^+$ / 1 mEq Na$^+$) = 19,090 mg Na$^+$

Total sodium chloride (NaCl) in sweat: (19,090 mg Na$^+$) × (1 g NaCl / 393 mg Na$^+$) = 48.6 g NaCl [b]

Intravenous replacement of excreted sweat sodium[c]

Given: 1 L of ½ normal saline intravenous fluid (0.45% NaCl) contains 77 mEq Na$^+$
   1 L of normal saline intravenous fluid (0.9% NaCl) contains 154 mEq Na$^+$
   1 L of 3% hypertonic saline intravenous fluid (3% NaCl) contains 513 mEq Na$^+$

Therefore, intravenous replacement[d] of 830 mEq Na$^+$ (19,090 mg Na$^+$) could require one of the following:

- 10.8 L of ½ normal saline (830/77 = 10.8): would induce iatrogenic illness
- 5.4 L of normal saline (830/154 = 5.4): likely would induce iatrogenic illness, unless athlete were considerably dehydrated
- 1.6 L of hypertonic saline (830/513 = 1.6): this is an appropriate solution to replace sodium loss shown above

Dietary replacement of excreted sweat sodium[c]

Given: 1 mEq Na$^+$ = 23 mg Na$^+$
   1 can of soup contains 85 to 107 mEq Na$^+$
   Canned tomato juice contains 66 mEq Na$^+$ per serving
   20 oz of a sports drink contains 6.5 to 21 mEq Na$^+$ · L^{-1}

Therefore, dietary replacement of 830 mEq Na$^+$ (19,090 mg Na$^+$) could be accomplished by one of the following:

- 7.8 to 9.8 cans of soup
- 12.6 servings of tomato juice
- 39.5 to 127.7 L of a sports drink

Clearly, a combination of salt-containing food items would be best, as any one option (as listed here, for example) would be inappropriate.

[a] Trained, heat-acclimatized athletes ordinarily exhibit sweat sodium concentrations between 20 and 40 mEq Na$^+$ · L^{-1}; this value of 83 mEq · L^{-1} (about 2 g NaCl · L^{-1}) is high but was observed by the author in a fit 17-year-old tennis player; this characteristic apparently predisposed him to recurrent heat cramps.
[b] A typical U.S. diet contains 8 to 13 g NaCl per day.
[c] Does not include urinary or other sodium losses.
[d] The amount of fluid consumed during the event and the present level of dehydration (i.e., deficit in liters), should be considered when selecting the volume and composition of an intravenous fluid.

Figure 5.1 Sample calculations of sodium (Na$^+$) and sodium chloride (NaCl) losses during exercise in a hot environment, and replacement of sodium via dietary and intravenous routes.
The author acknowledges the intellectual contributions of Lawrence E. Armstrong to this figure.

However, given that it takes 1 to 2 hours for peak blood levels to occur, a negative neuromotor effect may not be perceived or significantly reduce performance if quinine is orally administered toward the end of a game or match. Importantly, other potential adverse effects ranging from hypoglycemia to nausea, vomiting, disturbed vision, or even more serious toxic symptoms associated with higher dosages (37) should be considered before taking or administering quinine for heat cramps.

CONTINUATION OF ACTIVITY AND RETURN TO PLAY

If treatment is applied right away—that is, a salted solution is consumed at the onset of muscle twitches or before the heat cramps become too strong and debilitating—an athlete may be able to avert further progression of heat cramps and, in fact, resolve the symptoms, so that competition or training can continue without interruption. Even if more aggressive treatment (e.g., intravenous fluid) is required, an athlete often can return to play fairly soon. Football players can be treated at halftime and return for the second half of the game, seemingly without any detrimental effects. And tennis players often readily recover from cramps through appropriate postplay oral rehydration or intravenous treatment and perform quite well in a subsequent match the next day.

If the athlete exhibits accompanying symptoms implying that a more clinically significant and severe degree of heat strain has been incurred (e.g., diagnosed heat exhaustion), the athlete should not compete or train until given medical clearance. In the case of heat exhaustion, the athlete usually will need to wait 1 to 2 days before progressively and cautiously resuming intense training.

PREVENTION

Although recovery from heat cramps can be quite rapid, prevention is clearly more desirable. If an athlete is sweating at an extensive rate during exercise, it can be difficult (if not impossible) to concomitantly replace enough fluid to offset the ongoing high rate of water loss. Two primary factors limit fluid replacement. First, the volume of fluid in the stomach and the ratio of carbohydrate to volume primarily determine the rate at which fluid empties from the stomach. Greater volumes (e.g., by drinking 200 ml every 10 min) empty faster, but too much carbohydrate (e.g., via a >10% carbohydrate drink) can slow gastric (stomach) emptying. Thus, gastric emptying of water and many sports drinks (if they have 6% or less carbohydrate content) during exercise is seemingly not a limiting factor in maintaining hydration (38, 39). Second, a high sweating rate may limit an athlete's ability to fully replace water loss as it occurs. Athletes with very high sweating rates (>2 L · h^{-1}) and comparatively slower gastric emptying rates may not be able to replace fluid fast enough. Overconsumption of any fluid can lead to gastric discomfort and related performance decrements (40). In addition, it is unlikely that fluid absorption will be much greater than 2 liters per hour (38); consuming more than that would probably exacerbate feelings of discomfort, even if most of the fluid is readily leaving the stomach. Therefore, athletes who sweat heavily must drink plenty of fluids throughout the day so that they begin competition or training in a well-hydrated state. Simply monitoring certain indices such as urine volume and color can provide athletes with some insight as to whether they are well-hydrated prior to play (41, 42).

Rehydration after exercise can be equally challenging, particularly if the athlete must compete again soon on the same day. In some sports, players must return to the field or court in an hour or less, and this may not nearly be enough time to replace most of an extensive fluid deficit. Notably, to fully rehydrate after exercise, athletes may need to consume about 150% of the amount of fluid that is indicated by a loss of body weight due to sweating (43). For example, an athlete whose weight is down one pound at the end of play may need to drink another 24 ounces, to compensate for the obligatory urine production that occurs, even as the athlete is not yet completely rehydrated.

As important as consuming plenty of fluid is to rehydration, complete rehydration and restoration of the extracellular fluid compartment requires adequate sodium replacement as well (44, 45). Cramping athletes will often similarly report, "But I was drinking a lot of water! In fact, I was constantly urinating before the match [or game] began." This underscores that athletes cannot rely solely on frequent urination as an indicator of complete and adequate rehydration. These athletes often do not appreciate that they need to both drink plenty of fluids and replace their electrolyte losses. The appropriate amount of salt an athlete should consume is directly dependent on how much sodium and chloride were previously lost through sweating.

Practical Dietary Guidelines

In general, a diet for good health is also appropriate for good performance. However, adequate fluid intake should be a particular priority with most athletes. As mentioned, maintaining hydration during exercise and rehydration after exercise can be challenging. Sufficient carbohydrate intake is also important to support energy needs and may assist with restoring body water (46).

To prevent heat cramps, adequate electrolyte (namely, sodium and chloride) replacement should be emphasized. Athletes may add sodium to their diets by simply salting their meals or eating processed foods. A can of soup, for example, contains 1,950 to 2,450 milligrams of sodium chloride; a serving of canned tomato juice contains 1,525 milligrams of sodium chloride. The sodium content of some vegetables increases markedly when they are canned. Compare the sodium content of these raw and canned vegetables: green beans, 8 milligrams of sodium when raw vs. 536 milligrams of sodium when canned; carrots, 31 vs. 280 milligrams; green peas, 2 vs. 236 milligrams; and sweet potatoes, 24 vs. 48 milligrams. In comparison, fluid–electrolyte replacement beverages contain 150 to 275 milligrams per 20-ounce container. Also, certain foods ordinarily contain a large amount of salt (e.g., salted pretzels, tomato sauce, cheese, and pizza) and periodically may be emphasized in an athlete's diet.

Ideally, salt intake should be based on an individual's specific sweat and electrolyte losses for a given workload and environmental stress. Field and laboratory tests can provide insight and guidelines for the athlete, but many athletes are not likely to participate in such testing. Experimenting with known amounts of salt-containing beverages may be a more practical alternative (25). For example, consuming a well-mixed combination of 0.5 teaspoon of salt and 32 ounces of sports drink well before or some time after competition or training is an easy way to add additional fluid, carbohydrate, and 1,620 milligrams of sodium to one's diet. One or several of these mixtures consumed over the course of a day could be enough to make up the difference between sweat fluid and electrolyte losses and what the diet would otherwise provide. Athletes who are especially prone to heat cramps may need to consume sodium-containing fluids (e.g., 0.25 teaspoon of salt to 32 ounces of sports drink) during exercise and sport as well. Salt tablets can work too (see Treatment on page 96), as long as they are consumed with enough fluid. Even pickle juice is sometimes used in the prevention of heat cramps, but it is not a very palatable way to provide significant sodium or fluid volume. Any method for increasing salt intake should be tested prior to competition; a familiar routine is more likely to be better tolerated and more effective.

Acclimatization and Fitness

Athletes often mistakenly believe that being "used to," or acclimatized to, the heat and humidity reduces their susceptibility to the effects of extensive play in the heat. "Why am I cramping? I live in a hot climate!" represents the typical surprised response of athletes who thought that they were immune to such heat illnesses. It is true that well-conditioned and fully heat-acclimatized athletes often have higher sweat rate capacities and lower sweat sodium concentrations than do less fit or less heat-acclimatized people (47). For example, 5 to 30 millimoles of sodium per liter of sweat, which equates to only 115 to 690 milligrams of sodium per liter (48), can be typical for many athletes. However, extensive and repeated fluid losses via sweating will eventually lead to sizable sodium losses, even if the sweat sodium concentration is fairly low. If dietary salt intake is concomitantly low or insufficient, a significant sodium deficit (and potentially muscle cramping) likely will ensue. Heat acclimatization, in this situation, may not be enough to protect the athlete from heat cramps.

Sweat gland conservation of sodium can occur after 5 to 10 days of exercise in new environmental (heat and humidity) exposure. Unfortunately, such an early arrival is not always possible or practical for traveling athletes. Moreover, some athletes maintain relatively high sweat sodium levels, even when they are acclimatized to the heat. Thus, during and after the heat-acclimatization process, some athletes may still have to regularly consume extra salt in an effort to maximize rehydration and reduce the risk of heat cramps.

Recommendations

Following these recommendations can reduce the risk of developing heat cramps.

- Athletes should acclimatize themselves to the heat and humidity as much as possible prior to competition or extensive intense training in such environmental conditions.
- Athletes should drink plenty of fluids (water, juice, milk, sports drinks) and eat ample carbohydrates throughout the day.
- If prone to heat cramps, athletes should consume extra salt before, after, and possibly during training and competition.
- Certain athletes should have their sweat rate and sweat electrolyte losses measured, so that appropriate strategies can be implemented for restoring fluid and mineral balance.
- If heat cramps persist, athletes should see their doctor for potential causes related to medications, an underlying illness or metabolic disorder, or other predisposing factors.

Poor Treatment of Cramps Leads to Hyponatremia   Case Report 5.2

After playing (and winning!) a 4-plus-hour tennis match in the heat (>38°C; >100.4°F) and humidity of the Midwest summer, 17-year-old C.L. was extremely thirsty and began to experience diffuse heat cramps. During play, he drank mostly water and supplemental smaller amounts of a sports drink. C.L. was advised by tournament medical personnel to drink a lot of water and to go back to his hotel and rest. While in his room and after having consumed a considerable amount of water since leaving the tournament site, C.L. suffered a seizure and slipped into a coma. Not until more than 2 days later were his serum electrolytes stabilized (his initial blood work showed a serum sodium of 118 mmol · L^{-1}). C.L.'s discharge diagnosis was acute hyponatremia and coma.

C.L. was subsequently tested during 4 hours of competitive tennis in similar high heat and humidity. With a measured sweating rate of about 2.5 liters per hour and a sweat sodium concentration of almost 85 millimoles per liter, it was estimated that C.L. could have readily lost over 20,000 milligrams of sodium (50 g of salt) in the match preceding his episode of severe hyponatremia. Such significant sodium loss from profuse sweating, coupled with only water and low-sodium fluid intake during the match, likely predisposed C.L. for hyponatremia when he overhydrated after play.

This case emphasizes the importance of replacing sodium chloride and fluid when rehydrating, not only to prevent or resolve heat cramps, but also to reduce the risk of developing hyponatremia. This potentially dangerous condition is described in the next chapter.

SUMMARY

Although a variety of mineral deficiencies and physiological conditions are purported to cause muscle cramps, evidence suggests that extensive and repeated bouts of sweating and a consequent sodium deficit are the primary etiological factors associated with exertional heat cramps. Heat cramps often begin as subtle twitches or fasciculations in one or more localized areas and, unless treated, can rapidly progress to widespread debilitating muscle spasms that leave an athlete writhing in pain. Athletes can usually avoid heat cramps by taking sufficient preventive measures before and during competition. Appropriate sodium and fluid intake, based on one's specific losses, will ensure adequate rehydration and fluid distribution throughout the body and will greatly reduce the risk of exertional heat cramps, even in the most challenging of hot environments.

Chapter

Exertional Hyponatremia

Lawrence E. Armstrong, PhD, FACSM

Fluid Overload Leading to
Hospitalization and Disability

Case Report 6.1

A 49-year-old distance runner (G.M.) prepared for a nationally recognized marathon by competing in two half-marathons and three 42.2-kilometer (26.2-mile) marathons during an 18-month period. The event he had prepared for was held in the southern United States during the month of June. The prerace registration materials and the race Web site encouraged drinking fluids before, during, and after the race.

On the Friday and Saturday prior to the marathon, G.M. consumed foods that runners commonly eat before races: bread, pasta, salad, crackers, water, fruit, soda, and eggs. On Sunday, he completed the marathon in 4 hours and 22 minutes. Friends drove G.M. to the airport for the flight home, and he boarded the flight alone. During the flight, approximately 5 hours after the race, G.M. became ill and experienced a grand mal seizure in the aisle of the cabin. The pilots diverted the jet to a nearby airport and made an emergency landing. G.M. was rushed to the emergency department of a medical center, where he experienced two additional seizures while he was unconscious in the emergency department. G.M. had no previous history of nerve or muscle diseases, heart disease, lung disease, ulcers, or other medical conditions.

Upon admission to the hospital, G.M.'s serum sodium level of 129 mEq · L^{-1} was well below normal (i.e., 135-145 mEq · L^{-1}), prompting a physician to diagnose G.M. as having hyponatremia caused by excessive water retention. G.M.'s serum sodium level was corrected to 139 mEq · L^{-1} during the next 48 hours. A chest X ray showed diffuse pulmonary edema; G.M. was subsequently treated for respiratory distress syndrome with supplemental oxygen. A computed tomography scan indicated cerebral edema; this resolved with the treatment for hyponatremia. Pupilary reflexes were sluggish and he exhibited the "dolls eyes" sign (i.e., a reflexive movement of the eyes in a direction opposite that of the head).

G.M.'s condition was most serious throughout the initial 3 days of hospitalization, which included treatment for renal insufficiency. His initial serum creatine phosphokinase (CPK) level was 1,521; it rose to 7,000 on Monday and peaked at 17,000 IU on Wednesday, indicating muscle damage (i.e., rhabdomyolysis). The serum enzymes alanine aminotransferase (ALT) and aspartate aminotransferase (AST), indicative of liver damage, returned to normal spontaneously over several days.

On Thursday, all fluid and oxygen tubes were removed from G.M., and he was alert and oriented but drowsy. He remained in the intensive care unit through Friday and was discharged from the hospital on the ninth day; his CPK level had decreased to 5,000 IU. G.M. could not recall the number of water stations he had utilized nor the amount of water that he had consumed during the race. During the ensuing 18 months, G.M. learned that he was unable to mentally process information that had been routine prior to this experience. His cognitive function was damaged to the point that he no longer could perform his professional duties.

In most clinical textbooks, hyponatremia is not classified as a heat illness. However, because exercise in a hot environment stimulates a larger water (i.e., sweat) loss and fluid intake than does an identical workout in cold conditions, and because most cases occur during prolonged exercise in summer months, exertional hyponatremia (EH) is included as a heat-related illness in this book. Since 1989, many authorities have considered endurance exercise and a hot environment to be factors that predispose athletes and laborers to EH (1-9).

EH involves a serum or plasma sodium concentration of less than 130 mEq · L⁻¹ (some authors recognize 135 mEq · L⁻¹ as the definitive concentration), typically resulting from replacement of sweat losses with water or other hypotonic fluid. Severe cases may involve coma, pulmonary or cerebral edema, or death. Only this disorder and exertional heatstroke (chapter 3) are potentially fatal to otherwise healthy athletes, laborers, and military personnel.

COMMON SCENARIOS AND PREDISPOSING FACTORS

In recent years, physiologists, dieticians, physicians, and race directors have vigorously emphasized water consumption. At times, this advice has been interpreted by athletes and soldiers to mean "more is better, without limit." In view of case report 6.1, it is reasonable to wonder how many runners realize the potential of excessive fluid consumption to cause illness, hospitalization, cognitive impairment, or occupational disability. The apparent answer is only a few. In fact, some authorities have reported that the life-threatening potential of EH makes it the most serious mortality risk at some endurance events.

Reports of EH most often arise from military training (4, 6, 10), long hikes (2, 5), distance running events that are 42.2 kilometers (26.2 miles) or longer, and triathlons that last 7 to 17 hours (11-14). Hyponatremia can develop in as little as 4 hours (15, 16). There are no reports of hyponatremia in short events such as 5- or 10-kilometer races.

Chapter 8 presents other factors that may predispose active individuals to EH, including female sex, a slow running speed, a high sodium concentration in sweat, inappropriate hormone or renal responses (e.g., retention of fluid by the kidneys during fluid overload), and use of common nonsteroidal anti-inflammatory drugs such as aspirin or ibuprofen. See chapter 8 for more information on these topics.

CLINICAL AND PHYSIOLOGICAL FEATURES OF EXERTIONAL HYPONATREMIA

EH initially was observed among ultraendurance athletes after prolonged exercise of 7.0 to 16.9 hours (15). In recent years, however, drinking requirements have been emphasized strongly and internalized by competitors to the point that ingesting excessive volumes of hypotonic fluid during exercise has led to an increased incidence among marathon runners who run for 4.0 to 6.9 hours (14). Interviews with these competitors revealed that virtually all had attempted to drink "as much fluid as possible" during and after the race and that they previously had been unaware of the potential dangers of drinking too much fluid.

But athletes at certain events do not always consume excessive amounts of fluid. Pugh, Corbett, and Johnson published the findings of a classic field study in 1967 (17) that included the body weight changes, rectal temperatures, and sweat rates of marathon competitors. Environmental conditions during the race ranged from 22 to 23.5°C (71.6-74.3°F) and 52 to 58% relative humidity. The mean time to completion was 3 hours and 14 minutes, the mean sweat rate was 0.96 liters per hour, and the mean final rectal temperature was 39°C (102.2°F). Body weight loss during this event averaged 2.85 kilograms (6.3 lb; 4.4% loss of initial body weight). Figure 6.1 illustrates the distribution of total weight lost by 70 finishers during this 42.2-kilometer event. None of these runners gained weight. Similar findings

Marathon runners must consume fluids while racing, but it's important for them to optimize, not maximize, their fluid intake to help them avoid exertional hyponatremia.

were reported by Fallon et al. (18) after a cool weather 100-kilometer footrace (2-17°C, 35.6-62.6°F, ambient temperature) and by Speedy et al. (19) after a warm weather triathlon (21°C, 69.8 °F; 91% relative humidity). Mean body weight loss was 3.3 kilograms (4.2% of initial weight) in the former event and 2.5 kilograms (3.5%) in the latter. Zero (out of 7 total) and one (out of 15 total) individuals, respectively, gained weight during these events. Clearly, dehydration was prominent and EH was not of great concern in these competitions. Why then does EH occur at certain events and not others?

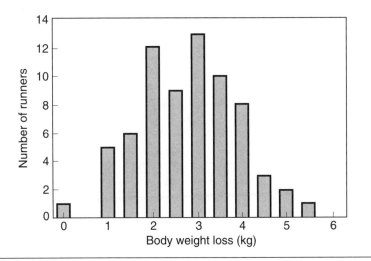

Figure 6.1 Body weight losses of 70 marathon runners who finished a 42.2-kilometer event in a mild environment (22-23.5°C; 71.6-74.3°F). No runner experienced exertional hyponatremia. Average water intake was 0.42 liters in 3.2 hours, resulting in an average body weight loss of 2.9 kilograms (5.2 %).

Adapted, by permission, from L G. Pugh et al., 1967, "Rectal temperatures, weight losses, and sweat rates in marathon running," *Journal of Applied Physiology* 23(3): 347-352.

Since its initial description in 1985 (20), hyponatremia increasingly has been recognized as a problem in ultraendurance events (e.g., the Hawaiian Ironman Triathlon and the Comrades Marathon in South Africa). Noakes (21) ascribed this phenomenon to the recent popularity of ultradistance events and the introduction of multiple watering stations that have allowed runners virtually unrestricted access to fluid. Coupled with the belief that they should drink as much as possible, these additional water stations allow athletes to voluntarily ingest large volumes of hypotonic fluid during the final miles of these races. Today, EH is recognized as one of the most common medical risks of ultradistance events (13, 22-24), even though the incidence is low (see chapter 2).

Definitions of hyponatremia differ among sources, but a serum sodium concentration below 136 mEq · L^{-1} is the point at which diagnosis becomes possible (14, 15, 25, 26). Mild hyponatremia generally is defined as a concentration between 130 and 135 mEq · L^{-1}, whereas patients with severe hyponatremia typically have serum sodium concentrations of <125 mEq · L^{-1}. Although 44 to 64% of ultramara-thon finishers require medical attention with serum levels below 130 mEq Na$^+$ · L^{-1} (16), most cases above this level are asymptomatic (2). For example, Speedy and associates (11) reported that all hyponatremic athletes with plasma sodium concentrations below 125 mEq · L^{-1} exhibited symptoms, and Backer and colleagues (2) noted that symptoms were obviously severe below this level.

Figure 6.2 presents the signs and symptoms that are most often observed in exercise-related hyponatremia. The prominent neurological manifestations include disorientation, depression, nausea, vomiting, muscular twitching, respiratory arrest, grand mal seizures, and coma (27). When severe hyponatremia is induced acutely by water intoxication—caused by ingesting large quantities of water or having kidney disorders that reduce urinary output—over just a few hours, mortality up to 67% (28) with substantial morbidity (usually brain damage) is possible in patients with renal disease and sedentary individuals. When exercise is a part of the etiology, the mortality rate apparently is less than 50%, based on published case reports. Although fatalities due to EH have occurred during marathon running (29) and military basic training (4), they are rare because athletes and military recruits perform in groups and have medical care available.

Change of mental status	Disorientation
Nausea	Confusion
Vomiting	Incoordination
Headache	Combative behavior
Dizziness	Physical exhaustion
Muscular twitching	Muscular weakness
Grand mal seizures	Cardiac arrest
Coma	Respiratory arrest
Somnolence	Pulmonary edema
Tingling	Cerebral edema

Figure 6.2 Signs and symptoms of acute exertional hyponatremia occurring in plasma as sodium lost at a rate of 0.5 mEq · L^{-1} · h^{-1}.
Data from 27, 29.

Theoretically, exertional hyponatremia may result from one of four plausible circumstances: (a) Sodium losses and the sweat rate during prolonged exercise are moderate or normal, but the sodium concentration and osmolality of plasma are reduced to critical levels when a large volume of hypotonic fluid is consumed and retained in the body (20, 30, 31); (b) extracellular fluid is diluted primarily because the sweat sodium concentration and the volume of sweat are both very large (16, 32, 33); (c) a large, inappropriate arginine vasopressin (AVP; also known as antidiuretic hormone, ADH) secretion induces excessive free water retention at the kidneys (15, 24, 34); and (d) extracellular fluid tonicity and volume are regulated inappropriately by a combination of the first three mechanisms mentioned (14, 15, 35, 36). In addition, numerous other predisposing factors (e.g., gender, volume of total body water) play a role; these factors are covered in chapter 8.

During the past decade it has become apparent that no single mechanism can explain all cases of EH (23). However, certain characteristics are common to the vast majority of EH experiences. First, water overload via sustained high rates of hypotonic fluid intake exists in virtually all instances of severe hyponatremia (24, 36). Second, the kidneys fail to excrete excess water, and hyponatremic athletes excrete scant concentrated urine, for reasons that are not completely understood (11, 24). Third, by definition, exercise is an etiological factor in all cases. During strenuous exercise, blood flow is diverted from the stomach, intestines, and other splanchnic organs to skeletal muscle and skin (37). This response results from (a) increased sympathetic nerve impulses to the kidneys that cause constriction of renal capillaries and (b) secretion of the hormones renin and angiotension that increase sodium and water reabsorption. These actions can reduce urine production by 20 to 60%, without invoking an inappropriate AVP secretion (16).

Both exercise and environmental heat stress cause urine output to decrease (38). Relative ischemia is possible in gastrointestinal tissue and may explain the well-known diarrheal distress suffered by long-distance runners who compete on hot days (39). Animal research also has shown that the intestine assumes the role of a temporary water and salt depot (40) when exposed to high ambient temperatures (33-40°C; 91.4-104°F). This response also may exist in cases of EH, in which ingested water remains in the gut for several hours before entering the circulation; but when exercise ceases, this water is abruptly absorbed into the bloodstream and the extracellular fluid is diluted (30). Noakes (41), in fact, stated that up to 60% of ingested fluid might remain in the intestinal lumen during 2 hours of exercise (42), when consumed at a rate of 900 milliliters per hour. Such a delayed entry into the blood ought to result in delayed signs and symptoms of hyponatremia as well. This phenomenon has been reported as delayed symptomatology in 4 out of 7 overhydrated hikers (2), 8 out of 16 water-loaded athletes (24, 30, 35), and one experimental test participant who drank excessively (15). Evidence from early studies of water intoxication in dogs (43) suggests that sodium losses into the intestine (to adjust the osmolality of chyme to have a tonicity equal to that of blood) may exacerbate the symptoms of hyponatremia. Convulsions were closely associated with the loss of sodium, via gastric secretions, from the extracellular compartment into the intestinal lumen. This highlights the importance of understanding fluid–electrolyte movements during the development of hyponatremia. The following paragraphs describe the findings of several human studies that have evaluated similar responses.

Fluid–Electrolyte Shifts and Symptomatology

Prior to 1985, the clinical literature regarding hyponatremia focused on psychological illnesses that involved aberrant, excessive drinking (i.e., polydipsia) in psychiatric patients (44). Symptomatic hyponatremia develops in about half of these water drinkers because of water intoxication, secondary to chronic schizophrenia (44, 45). Much of our understanding of fluid–electrolyte shifts during the development of hyponatremia originated from observations of animals and sedentary, long-term psychiatric patients. However, because these cases did not involve the critical stress of exercise, these studies must be interpreted with caution to avoid unwarranted inferences. Table 6.1 presents a compilation of statements by various authors about fluid–electrolyte factors of nonexertional (sedentary) acute hyponatremia in various scenarios.

Table 6.1
Observations Regarding Sedentary Cases[a] of Acute Symptomatic Hyponatremia in Humans and Animals

Scenario	Observations
A healthy person who has consumed an excess of water or has received an excess of intravenous fluid. Pituitary, kidney, and adrenal function are normal.	• It is unusual and difficult for a normal person to consume enough water to induce water intoxication (46, 47, 25). • Clinically, it is almost impossible to induce hyponatremia by the loss of sodium alone (48). • The extracellular fluid becomes dilute and its osmotic pressure becomes less than that inside cells. As a result, water flows into cells, causing them to swell (47). • Serum sodium level decreases below 135 mEq · L^{-1}; other serum electrolytes also decrease (47).

Scenario	Observations
	• Acute hyponatremia develops at a rate of 0.5 mEq $\cdot$ L^{-1} $\cdot$ h^{-1} and typically involves fluid overload. Chronic hyponatremia occurs over several days at a slower rate, exhibits different and less severe symptoms, and involves a lower morbidity and mortality. Numerous diseases, and exposure to a hot environment while consuming a low-salt diet, may cause chronic hyponatremia (28).
Hospitalized adults.	Hyponatremia is the most common electrolyte abnormality in hospitals (28).
Postoperative patients with hyponatremia.	The majority of postoperative patients exhibit elevated blood levels of AVP, likely caused by a reduced ECV. ECV contraction provides a strong stimulus for AVP release (49).
Self-induced water intoxication in psychiatric patients (plasma sodium <125 mEq $\cdot$ L^{-1}).	• Obvious neurological symptoms occur (e.g., nausea, vomiting, muscular twitching, grand mal seizure, coma) because of brain swelling (i.e., cerebral edema). Postmortem autopsies show obliteration, flattening, and herniation of brain structures (28, 50, 51). • Rare reports exist of fatalities in psychiatric polydipsia patients with normal renal free water clearance (51, 52).
Acute hyponatremia induced in experimental animals via fluid overload.	• The extracellular volume expands; more swelling occurs in muscle and liver than in the brain (28, 53). • An osmotic equilibrium between the brain and plasma requires more than 2 h to achieve (48). • Acute cases, developing in 1 to 4 h, involve greater brain water content than do chronic cases (53). • When acute hyponatremia remains for more than 3 to 4 h, the brain extrudes intracellular electrolytes (Na^+, Cl^-, K^+) and small organic molecules (osmolytes) to minimize brain swelling (48, 54).

[a]All cases involved fluid overload while at rest.

mEq $\cdot$ L^{-1} = milliequivalents per liter; ECV = extracellular volume; AVP = arginine vasopressin; h = hour; Na^+ = sodium; Cl^- = chloride; K^+ = potassium.

Although none of the descriptions in table 6.1 refer to EH, it is reasonable to expect similarities between some forms of sedentary and exertional hyponatremia because they likely share a common cause—fluid overload. Thus, table 6.2 provides a compilation of statements that describe fluid–electrolyte factors associated with prolonged, low-intensity EH. A comparison of table 6.1 and table 6.2 indicates that

Table 6.2

Observations Regarding Human Cases[a] of Acute Symptomatic Exertional Hyponatremia (EH)

Scenario	Observations
Ultradistance triathletes and runners who exercise for more than 7 h	• No single mechanism explains all cases of EH. Multiple factors are responsible for this disorder (15, 23). • Symptomatic EH usually develops in the presence of gross fluid overload (14, 15, 24, 41, 43). • Smaller body size, smaller total body water, female gender, longer time on the course, and use of NSAIDs have been implicated in the etiology of EH (11, 14, 16, 24, 29, 55). • Symptomatic EH likely involves an acute expansion of the ECV and PV, with dilution of sodium in both. The duration and dissolution of this expansion are unknown (15, 26, 36, 56). • An inverse statistical correlation exists between body weight change and postrace serum Na^+ concentration; as the former increases, the latter decreases (36). • Fluid overload can result from fluid accumulation in a physiological third space (e.g., intestine) with subsequent Na^+ movement into that fluid, impaired renal excretion of a fluid load, or fluid intake that exceeds the maximal capacity for urine production (15, 16, 36, 41). • EH developed in five competitors who consumed fluid at a modest rate (421-766 ml · h^{-1}) during a triathlon (36). • Renal function was not impaired in runners with EH, but they neither excreted excess water nor adequately controlled urinary Na^+ loss (24). • EH may result in death, but this is rare (4, 29). • Most data do not support inappropriately high AVP levels in cases of EH (36, 55).

Scenario	Observations
Healthy adults, during prolonged walking or hiking	• One young male consumed fluid excessively (10.3 L · 4 h⁻¹) while walking. His plasma AVP was normal up to 4 h, but increased markedly (460%) between 4 and 7 h; this increase coincided with a decrease of urine output to 0 ml · h⁻¹ (15).
	• Exercise can decrease urine production rate by 20 to 60%, because of increased sympathetic nerve activity to the kidneys and increased hormone secretion (e.g., renin and angiotensin) that increases reabsorption of Na^+ and water (16).
	• Sodium losses in sweat and urine were normal and served only to exacerbate EH (15, 24).
	• A preexercise serum Na^+ level that is "low normal" (e.g., 134 mEq · L⁻¹) may predispose a person to EH (15, 24).

[a] All cases involved fluid overload during prolonged, low-intensity exercise.

Na^+ = sodium; NSAID = nonsteroidal anti-inflammatory drug; AVP = arginine vasopressin; ECV = extracellular volume; PV = plasma volume; ml · h⁻¹ = milliliters per hour; mEq · L⁻¹ = milliequivalents per liter.

sedentary hyponatremia and EH share many characteristics. This suggests that investigators may uncover clues in the sedentary hyponatremia literature that clarify the etiology and physiological mechanisms underlying EH. For example, table 6.1 includes information derived from animal experiments, psychiatric patients, and postoperative patients that cannot be found in the literature regarding EH.

Three Types of Acute Hyponatremia

Under normal conditions, the human body regulates plasma sodium and osmolality by controlling the whole-body balance of free water. Water balance involves both the control of intake and the renal excretion of water that is free of solutes (i.e., free water). This is achieved through the sensation of thirst, inhibition of the release of AVP from the brain, and decreased reabsorption of water (i.e., increased free water loss) in kidney nephrons. In contrast to this osmoregulation via water balance, extracellular volume regulation occurs via the control of whole-body sodium balance.

Low serum osmolality (i.e., hypotonicity) commonly coexists with low serum sodium and is named **hypotonic hyponatremia.** Once this condition is identified, further evaluation should be based mainly on the clinical assessment of the size of the extracellular space. Because sodium concentration is a primary determinant of extracellular fluid volume (ECV), the three recognized forms of acute hypotonic hyponatremia are classified in terms of the total body sodium level (25). **Hypervolemic hyponatremia** involves an excess of total body water, with an expanded ECV. **Euvolemic hyponatremia** occurs when there is an excess of total body water with either normal or slightly reduced whole-body sodium; in this case, the ECV is normal. **Hypovolemic hyponatremia** involves a shrinking of ECV with deficits of total body sodium and water. Table 6.3 lists characteristics of these three forms of hypotonic hyponatremia. Although treatment of these disorders is described later in this chapter, it is noteworthy that hyponatremic patients with decreased ECV should be treated via water restriction. Administration of hypertonic saline and diuretics may be necessary to subsequently reduce intracellular water (see Treatment on page 129).

FLUID–ELECTROLYTE STATUS OF ATHLETES WITH EXERTIONAL HYPONATREMIA

Understanding the three acute forms of hypotonic hyponatremia gives us insight into the fluid–electrolyte status of race finishers who experience EH. For example, see figure 6.3, which shows the data presented by Speedy et al. (11) regarding an ultradistance triathlon in New Zealand. Although those authors and subsequent commentators did not publish the following analysis, figure 6.3 provides a powerful tool to understand the nature of the various fluid–electrolyte perturbations. The vertical dashed line in this figure represents the boundary between runners who gained body weight and those who did not. The horizontal shaded zone depicts those triathletes who completed the race without experiencing hyponatremia (defined by Speedy and colleagues as plasma sodium >135 mEq $\cdot$ L^{-1}). In all, 330 competitors are represented. Eleven were hyponatremic with a weight gain (lower right), 47 triathletes were hyponatremic with a loss of body weight (lower left), and the vast majority (272 or 87%) were not hyponatremic. The line of best fit (i.e., linear regression) demonstrates a significant inverse relationship ($p < .0001$) between postrace plasma sodium concentration and percent change in body weight. In general, competitors with the lower plasma sodium concentrations either gained weight or had the smallest weight loss. This supports the concept that excessive water consumption is a component of the etiological mechanism of EH, especially considering that these endurance athletes likely lost between 0.8 and 1.0 liters of sweat per hour (17-19, 58, 59) and about 9 to 12 liters of sweat during this 12.4-hour event (11, 16, 59).

Table 6.3

Fluid–Electrolyte Characteristics of the Three Types of Acute Hypotonic Hyponatremia

	Hypervolemic hyponatremia	Euvolemic hyponatremia	Hypovolemic hyponatremia
Water and sodium balance	Water excess with elevated body sodium	Water excess with normal or slightly reduced body sodium	Primary sodium deficiency
Mechanism causing hypotonic hyponatremia[a]	Dilution	Dilution	Sodium loss
Water excess	Present	Present	Absent
Whole-body sodium	Excessive	Normal or slightly reduced	Deficit exists
Whole-body dehydration	Absent	Absent	Present
Extracellular fluid volume	Expanded, caused by water excess	Normal or slightly expanded	Contracted, caused by sodium loss
Clinical edema	Present	Absent	Absent
Intracellular fluid volume	Expanded	Expanded	Expanded
Etiology	Water excess occurs by oral ingestion (e.g., exertional hyponatremia), renal impairment or failure, excessive AVP secretion, congestive heart failure, liver cirrhosis, intravenous therapy, or surgical procedures.	Water excess occurs via drugs that inhibit renal water excretion (e.g., NSAIDs) or excessive and inappropriate AVP secretion.	Primary sodium deficiency occurs via sweat and renal losses, diarrhea, or vomiting replaced with water; diuretic use; or adrenal insufficiency. Sodium also may leave the extracellular space and accumulate in a "third space" (e.g., the intestines).

[a]Hypotonic (i.e., hypoosmotic) hyponatremia refers to low osmolality and low sodium concentrations in plasma or serum.

AVP = arginine vasopressin; NSAID = nonsteroidal anti-inflammatory drug.

Data from 27, 47, 57.

Figure 6.3 can be understood more thoroughly if the reader mentally superimposes figure 6.4 over figure 6.3. The lower left and lower right quadrants in figure 6.4 represent hyponatremic states; their counterparts among race finishers appear in the lower half of figure 6.3. This comparative process allows the reader to conceptualize the relative incidence of dehydration (i.e., weight loss), fluid excess (i.e., weight gain), and hyponatremia.

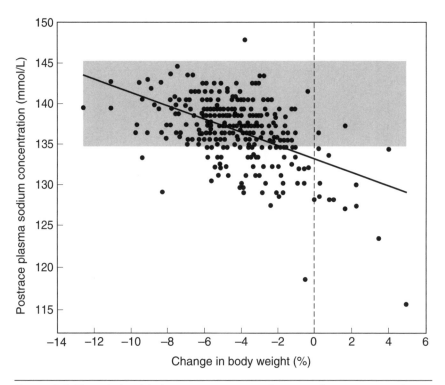

Figure 6.3 The relationship between postrace plasma sodium concentration (mmol · L⁻¹) and percentage change of body weight in 330 triathletes who participated in the 1997 New Zealand Ironman Triathlon (3.8-km swim, 180-km bicycle ride, 42.4-km run). The gray zone represents the range of normal plasma sodium concentrations.

Reprinted, by permission, from D.B. Speedy, 1999, "Hyponatremia in ultradistance triathletes," *Medicine and Science in Sports and Exercise* 31(6): 809-815.

Some hyponatremic triathletes (<135 mEq · L⁻¹) gained body weight (see figures 6.3 and 6.4, lower right quadrants), whereas others lost body weight (see the lower left quadrants). These states suggest that individuals who gained weight had a fluid excess, whereas

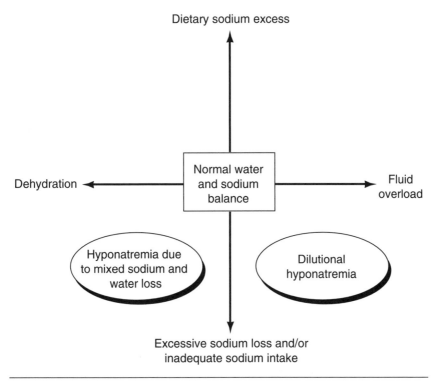

Figure 6.4 Possible changes in normal water and sodium balance. The lower left and lower right quadrants represent two different etiologies, both leading to exertional hyponatremia.

those who lost weight experienced a total body water reduction (i.e., dehydration). In terms of the extracellular fluid space, those with weight gain and fluid excess can be categorized as hypervolemic or euvolemic (see table 6.3) and those with weight loss and dehydration can be categorized as hypovolemic. Further, those who lost weight likely did so because their sweat sodium loss was large and the extracellular space contracted. Other fluid–electrolyte characteristics of these distinct groups may be found in table 6.3. Determination of these fluid–electrolyte states (i.e., sodium loss or fluid overload) should guide the clinical management of EH.

Plasma Volume Changes

In the 1999 New Zealand Ironman Triathlon study previously described (11), maintaining a normal plasma sodium level (137 mEq · L⁻¹) required a mean body weight loss of about 4% (figure 6.3). This seemingly curious fact can be explained by the body's tendency to reduce extracellular fluid volume subsequent to sodium loss. When facing a sodium deficit caused by sweat loss during prolonged exercise (e.g., 153 mEq sodium lost in 15.6 h) (24), extracellular dehydration protects against hyponatremia by reducing the amount of solvent (water) that sodium is dissolved in (41). The degree of contraction of the extracellular fluid is proportional to the magnitude of the sodium chloride deficit (60) during exercise. This phenomenon was lucidly demonstrated by Romero and colleagues (61), who fed sedentary adults very low sodium diets (<5 mEq · day^{-1}) for 6 days. At the end of this period, the mean plasma sodium level fell by only 1 mEq · L^{-1}, because extracellular volume decreased by 0.32 liters (6 %). However, this is not necessarily the response that occurs during ultraendurance events.

Reports of plasma volume changes during prolonged exercise only are reliable when pre- and postevent hematocrit (hct) and hemoglobin (hg) measurements are compared (62). For example, a recent report assumed that plasma volume contraction occurred (14), but this report should be interpreted cautiously because hematological values were not reported.

Table 6.4 presents the results from five studies that measured hematologic variables during the development of EH (15, 26, 56, 63, 64). Six features of this table should be noted. First, the plasma volume change of the three symptomatic EH cases (see symbol S in column 2 of table 6.4) ranged from +7.6% to +24.8%, all indicating expansion. However, the asymptomatic groups exhibited no consistent volume changes (–10.7% to +10.6%). Second, larger decreases in plasma sodium concentration resulted in larger plasma volume expansion. When the final plasma sodium increased, plasma volume decreased. Third, the two symptomatic ultramarathon runners had much higher plasma volume increases than did the symptomatic laboratory test subjects. This likely occurred because the laboratory volunteer began with a "low normal" plasma sodium level (134 mEq · L^{-1}) that was below the initial values of any other individuals; this extracellular dilution suggests an expanded plasma volume prior to exercise. Indeed, the subject's case history (15) indicated that he had consumed a large quantity of water on the previous day.

Fourth, numerous factors influence plasma volume shifts, including posture and exercise. The research of Pivarnik, Leeds, and Wilkerson (65) demonstrated that the intensity of upright exercise is especially important. During 1 hour of treadmill exercise, performed at three different speeds, plasma volume either increased or decreased in trained endurance athletes. At 37, 56, and 74% of maximal aerobic power ($\dot{V}O_2max$), plasma volume changed +3.3%, –0.5%, and –3.7%, respectively. Mean corpuscular hemoglobin concentration (MCHC) was quite similar across all groups. Interestingly, the exercise intensities of all subjects in table 6.4 were similar, at approximately 40 to 55% of $\dot{V}O_2$max, despite differences in their experiences (e.g., ultramarathon, field march, treadmill walk). Within this range of intensities, the research of Pivarnik, Leeds, and Wilkerson (65) suggests that plasma volume should slightly increase or remain stable if a large quantity of water was not ingested. Thus, upright exercise undoubtedly influenced the plasma volume changes of everyone described in table 6.4 to a minor and unknown extent, but fluid overload likely influenced plasma volume to a greater extent.

Fifth, it is possible for endurance athletes to experience a plasma volume decrease (63, 64), but this likely occurs only when the rate of fluid intake is small (e.g., <0.38 L · h^{-1}). Sixth, the three symptomatic individuals (S in table 6.4) consumed fluid at an average rate of 1.13 liters per hour, whereas the four asymptomatic comparison groups (A in table 6.4) drank at a mean rate of 0.48 liters per hour. Apparently, the total volume of fluid consumed was less important than the rate at which fluid was consumed. The data of Galun et al. (56) is especially noteworthy in this matter, in that the 17 males who marched over ground consumed 10.4 liters, the largest volume in this table. Yet, because this volume was consumed over 22 hours with meals (2,500-4,200 kcal, 1.5-4.1 g Na^+), renal handling of fluids and electrolytes was adequate to maintain mean plasma sodium at 135 mEq · L^{-1}. The subjects' relatively slow rate of intake (0.47 L · h^{-1}) apparently helped limit symptoms and maintain a normal plasma Na^+. In further support of this concept, Speedy et al. (66) recently demonstrated that a high rate of fluid intake (3.4 L · 2 h^{-1}) induced symptoms of illness in resting test subjects whose final plasma sodium concentration indicated mild hyponatremia (132-133 mEq · L^{-1}).

Intracellular and Extracellular Fluid Movements

The **mean corpuscular hemoglobin concentration** (MCHC) is an index of red blood cell swelling or shrinking; see the footnote in table 6.4 for the calculation. The MCHC of the symptomatic individuals (symbol S in column 2) decreased (–0.4 to –2.5), whereas the MCHC of the four asymptomatic groups (symbol A in column 2) either decreased or increased (–1.0 to +1.1). The data in table 6.5, which shows calculated fluid movements in normonatremia, EH, and resting hyponatremia, do not distinguish symptomatic and asymptomatic individuals. Although MCHC is often used as an index of intracellular swelling (i.e., the fluid inside red blood cells is assumed to represent the intracellular environment of other cells), this may not be an accurate representation of the condition of all of the body's cells. The normal range of MCHC values is 32 to 38%

Table 6.4

Plasma Volume Change, MCHC, and Other Relevant Variables During the Development of Symptomatic Acute Exertional Hyponatremia (S) and in Four Asymptomatic Comparison Groups (A)

Subjects, number, duration of exercise	Clinical status	Plasma Na⁺ (mEq · L⁻¹) Initial	Final	Plasma volume change (%)ᵃ	MCHC (g HG · 100 ml RBC⁻¹) Initial	Final	Total fluid intake (L)	Rate of fluid intake (L · h⁻¹)	Reference
Ultramarathon runner, n = 1, 13.3 h	S	140	131	+24.8	33.1	32.7	9.8	0.74	26
Ultramarathon runner, n = 1, 12 h	S	142	130	+15.7	33.5	32.8	9.2	0.77	26
Laboratory volunteer, intermittent walking on a treadmill, n = 1, 4 h	S	134	126	+7.6	34.7	32.2	7.6	1.90	15
Males marching over ground, n = 17, 22 h	A	142	135	+10.6	32.3	32.9	10.4	0.47	56
Laboratory volunteers, intermittent walking on a treadmill, n = 9, 4 h	A	140	136	+8.4	33.7	33.6	2.5	0.63	15
Laboratory volunteers running on a treadmill, n = 6, 3.3 h	A	144	147	-0.7	33.3	34.4	1.5	0.45	64
Ultramarathon runners, n = 170, 8.6 h	A	139	142	-10.7	33.6	32.6	3.3	0.38	63

ᵃCalculated by the authors from preexercise and postexercise hematocrit and hemoglobin, using the Dill and Costill (62) method.

S = symptomatic acute hyponatremia; A = asymptomatic individuals (≥135 mEq · L⁻¹); HG = hemoglobin; RBC = red blood cells; Na⁺ = sodium; MCHC = mean corpuscular hemoglobin concentration, an index that shows the relationship between the size of RBC and their HG content (MCHC = [HG/hematocrit] × 100); mEq · L⁻¹ = milliequivalents per liter; L · h⁻¹ = liters per hour.

(67), suggesting that none of the individuals in table 6.5 were abnormal. Although this suggests, albeit falsely, that there is no measurable change in intracellular water for symptomatic (versus asymptomatic) individuals, it likely reflects the inability of MCHC calculations to represent fluid movements into brain and other body tissues. Therefore, we discourage the use of MCHC to estimate the intracellular water volume changes of EH patients.

Simple calculations, however, can provide estimates of water movement between the intracellular and extracellular fluid compartments if the preexercise and postexercise plasma osmolality, net fluid balance, and body weight are known. Using the method of Renkin and Cala (68), table 6.5 presents such calculations for nine exercising normonatremic test subjects (column 2), a single test subject who experienced EH during exercise (column 3), and a hypothetical 70-kilogram (154.4-lb) male who experienced hyponatremia after consuming 5.0 liters of pure water (column 4). Fluid balance in these adults involved a net fluid loss of 1.2 liters, a net fluid excess of 2.8 liters, and a net fluid excess of 5.0 liters, respectively (see Scenario in the first column). The nine exercising test subjects (column 2) lost an equal volume of water (0.6 L) from both the intracellular fluid (ICF) and extracellular fluid (ECF) compartments (see final two rows). The test subject with EH (column 3) gained 3.1 liters of intracellular water and lost 0.3 liters of extracellular water during 4 hours of exercise. The hypothetical male (column 4) gained water in both the ICF (4.5 L) and ECF (0.5 L) compartments. Contradictory to the typical MCHC calculations, these latter two individuals exemplify the primary etiological danger of EH. Central nervous system (CNS) manifestations of this disorder (e.g., mental confusion, convulsions, coma) result from increased brain water and require aggressive medical treatment. Convulsions and coma represent a life-threatening emergency and should be treated with appropriate intravenous solutions, to create a concentration gradient that rapidly causes water to move out of the brain (see Treatment on page 129).

WHAT VOLUME OF WATER IS REQUIRED TO INDUCE EXERTIONAL HYPONATREMIA?

During the same era that Hiller (32) and O'Toole (33) recognized hyponatremia in Ironman triathletes, Timothy Noakes, MD, began a decade-long campaign to warn physicians, athletes, and race directors about the perils of water intoxication (13, 20, 21) resulting from voluntary overhydration. Although EH occurs today in few competitors (see chapter 2), it remains a serious health risk for some and can result in death or permanent disability. Noakes's most recent admonition involves the improper use of intravenous fluids to

Table 6.5

Calculated Fluid Movements Between the Intracellular and Extracellular Compartments in Normonatremia, Exertional Hyponatremia (EH), and Resting Hyponatremia

	Laboratory subjects (n = 9) with normal plasma [Na+], no EH	Laboratory test subject (n = 1) with EH	Hypothetical 70-kg resting male (n = 1) with hyponatremia
Pre plasma [Na+] (mEq · L⁻¹)	140[a]	134[a]	140
Pre TBW (L)[b]	46.7	49.3	42.0
Pre ICV (L)[c]	31.3	33.0	28.1
Pre ECV (L)[c]	15.4	16.3	13.9
Pre ECF and ICF osmolality (mOsm/kg)	287[a]	282[a]	290
Pre ICF osmoles (total mOsm)	31.3 × 287 = 8983	33 × 282 = 9306	28.1 × 290 = 8149
Scenario	Net fluid loss of 1.2 L / 4 h exercise	Net fluid excess of 2.8 L / 4 h exercise	Consumed 5.0 L of pure water, at rest
Post TBW (L)	45.5	52.1	47.0

Post plasma [Na$^+$] (mEq · L^{-1})	136[a]	122[a]	126
Post ECF osmolality (mOsm/kg)	280[a]	254[a]	259
Post ECF osmoles (total mOsm)	15.4 × 280 = 4312	16.3 × 254 = 4140	13.9 × 259 = 3600
Equilibrated ECF and ICF osmolality (mOsm/kg)[d]	(8983 + 4312) / 45.5 = 292	(9306 + 4140) / 52.1 = 258	(8149 + 3600) / 47.0 = 250
Calculated ICV (L)[d]	8983 / 292 = 30.7	9306 / 258 = 36.1	8149 / 250 = 32.6
Calculated ECV (L)[d]	4312 / 292 = 14.8	4140 / 258 = 16.0	3600 / 250 = 14.4
Calculated ICV change (L)[d]	-0.6	+3.1	+4.5
Calculated ECV change (L)[d]	-0.6	-0.3	+0.5

[a] Measurements made by Armstrong et al. (15); compare to table 6.4. [b] Calculated as 60% of body mass (kg). [c] Calculated as ICV = 67% of TBW, ECV = 33% of TBW. [d] Using the method of Renkin and Cala (68).

[Na$^+$] = sodium concentration; TBW = total body water; ECV = extracellular volume; ICV = intracellular volume; ECF = extracellular fluid; ICV = intracellular fluid; mEq · L^{-1} = milliequivalent per liter; mOsm/kg = milliosmoles per kilogram of water; [ ] = concentration of.

treat heat exhaustion and collapsed race finishers (58). He provides reasonable arguments that iatrogenic illness (i.e., illness induced by treatment), resulting from improper administration of fluid therapy, exacerbates EH.

Further, during the past three decades, water consumption has been emphasized strongly by nutritionists and fitness authorities. Military units and athletic organizations schedule rest periods during training, with the primary purpose of encouraging drinking. Indeed, since 1989 the military has experienced a rise in the incidence of EH that roughly coincides with an Army-wide emphasis on increased water intake during training in hot environments (W.M. Latzka, PhD, U.S. Army Research Institute of Environmental Medicine, Natick, MA, personal communication, June 2001). In Western societies, bottled water has become a symbol of fitness knowledge and good health, to the point that the public spends millions of dollars on spring and mineral water each year.

Thus it is reasonable to ask, "How much is too much?" Several research groups have measured fluid balance during endurance events, and other investigative teams have published dietary records of exercise enthusiasts who did not experience EH. These reports provide valuable insights and answers to this question. Table 6.6 presents fluid intake, estimated sweat losses, and weight loss data from such studies, ranked from the greatest (9.21 L) to the smallest (0.42 L) fluid intake. Three conclusions arise from this table. First, total fluid intake and the rate of fluid consumption increased with increasing duration of these events. This is likely because participants in such events exercise at a comfortable pace (e.g., 45-55% $\dot{V}O_2$max) and stop to drink heartily. Second, sweat rates were similar (0.8-1.0 L · h^{-1}) in a wide range of environments (2-28°C; 35.6-82.4°F) despite a wide range of fluid intakes (0.13-0.75 L · h^{-1}). Third, subjects who experience no signs or symptoms of EH safely experience mean body weight losses between 2.0 and 3.3 kilograms. These safe losses represent a range of 3.1 to 5.2% of initial body weight. Clearly, this is a much different fluid profile than individuals with fluid overload who retain several liters of fluid (see figure 6.3, lower right quadrant). This suggests that endurance athletes may safely consume fluids at a rate of about 0.74 or 0.75 liters per hour (see table 6.6) without experiencing symptoms of EH. However, the observations of Speedy et al. (26) clarify this statement, in that two ultramarathon runners developed symptomatic EH (plasma sodium levels of 130 and 131 mEq · L^{-1}) after drinking at rates of 0.77 and 0.74 liters per hour for 12.0 and 13.3 hours, respectively. Thus, it appears that fluid intakes of about 0.70 liters per hour represent an approximate threshold in the development of EH signs and symptoms.

Table 6.6

Fluid–Electrolyte Balance During Prolonged Endurance Exercise, Performed by Individuals With No Symptoms of Exertional Hyponatremia

Event, ambient temperature, number of participants	Mean duration (h)	Plasma Na⁺ (mEq · L⁻¹)		Incidence of exertional hyponatremia[a]	Total fluid intake (L)	Fluid intake rate (L · h⁻¹)	Sweat rate (L · h⁻¹)	Body weight change[b]		Reference
		Initial	Final					kg	%	
246-km triathlon at 21°C (69.8°F), n = 18	12.3	140	138	1 out of 18 runners	9.21[c]	0.75	0.8-1.0	-2.5	-3.5	19
160-km triathlon at 28°C (82.4°F), n = 13	10.3	—	—	None reported	7.62[c]	0.74	0.94	-3.2	-4.6	60
100-km foot race at 2-17°C (35.6-62.6°F), n = 7	10.5	—	—	None reported	5.70[c]	0.54	0.86	-3.3	-4.3	18
67-km foot race at 7-11°C (44.6-51.8°F), n = 170	8.6	139	142	No	3.29[c]	0.38	—	-2.4 ♂ -2.3 ♀	-3.3 -4.0	63
42-km laboratory treadmill run at 19°C (66.2°F), n = 6	3.3	144	147	No	1.46	0.45	—	-2.0	-3.1	64
42-km foot race at 22-23.5°C (71.6-74.3°F), n = 56	3.2	—	—	None reported	0.42[c]	0.13	0.96	-2.9	-5.2	17

[a]Plasma Na⁺ concentration <130 mEq · L⁻¹. [b]Including sweat, urine, or respiratory water loss (varied among studies). [c]Estimated or verbally reported.

Na⁺ = sodium; mEq · L⁻¹ = milliequivalents per liter .

125

Body size, total body water, beverage composition, sweat rate, and sweat sodium concentration also must be considered. Larger runners secrete a larger volume of sweat than do their smaller counterparts. Concentrated sweat (e.g., 60 mEq $Na^+ \cdot L^{-1}$) also serves to exaggerate the development of hyponatremia. Table 6.7 presents calculations that illustrate the interactions of these variables. Three hypothetical ultramarathon (90-km) runners are depicted, each with different sweat sodium concentrations, ranging from 20 to 60 mEq $\cdot L^{-1}$. In these theoretical examples, sweat volume was assumed to be matched by an equal volume of fluid consumption, during 9 hours of running. The resultant sodium losses and plasma sodium concentrations appear in table 6.7, below the values for sweat loss. The final row presents the excess amount of water required in each of the nine situations, to dilute plasma sodium to 120 mEq $\cdot L^{-1}$ during the race. For the two smaller runners, 2.2 to 4.2 (runner A) and 3.0 to 5.9 (runner B) liters of water, in excess of sweat volume, were necessary to induce a plasma sodium level of 120 mEq $\cdot L^{-1}$ (i.e., the threshold severe symptomatic EH). Such volumes are commonly consumed during prolonged endurance events (see tables 6.4 and 6.6). During a 9-hour race, these volumes represent excess consumption at rates of only 0.2 to 0.5 and 0.3 to 0.7 liters per hour; they represent total consumption rates of 0.9 to 1.1 and 1.3 to 1.6 liters per hour. Such rates are not common during prolonged exercise (compare to tables 6.4 and 6.6).

IS RHABDOMYOLYSIS ASSOCIATED WITH EXERTIONAL HYPONATREMIA?

As described in chapter 3, rhabdomyolysis involves skeletal muscle injury that alters the integrity of cell membranes to the point that a cell's contents escape into the extracellular fluid. The most common etiological factors in Western countries are alcohol abuse, drug administration, generalized seizures, localized disease, muscle compression, and prolonged strenuous exercise (69-71). Of particular interest, studies by Knochel and colleagues (72, 73) demonstrated that (a) young men in football or military training are particularly susceptible to environmental heat injury in hot weather, (b) this state may be accompanied by rhabdomyolysis, and (c) training under such conditions can lead to serious whole-body potassium depletion. These findings suggest that potassium depletion may increase susceptibility to rhabdomyolysis and that these disorders can occur during exercise–heat stress.

The link between EH and rhabdomyolysis is illustrated by a case report involving a 19-year-old man who hiked strenuously for several hours in a warm–humid environment and consumed several liters of tap water (74). Initial laboratory tests verified acute

Table 6.7

Calculated Water Excess That Is Required to Dilute Plasma Na$^+$ to 120 meq · L^{-1} in Theoretical Ultramarathon Competitors With Different Body Masses

	Runner A			Runner B			Runner C		
Body mass (kg)	50			70			90		
Body mass (lb)	111			155			199		
Total body water (L)	31.5			44.0			56.6		
Initial plasma Na$^+$ concentration (mEq · L^{-1})	140.0			140.0			140.0		
Initial extracellular fluid volume (L)	12.5			17.5			22.5		
Initial extracellular Na$^+$ content (total mEq)	1750			2450			3151		
Volume of pure water consumed (L · 9 h^{-1})	6.1			8.6			11.1		
Sweat loss (L · 9 h^{-1})[a]	6.1			8.6			11.1		
Three likely sweat Na$^+$ concentrations (mEq · L^{-1})[b]	20	40	60	20	40	60	20	40	60
Resultant Na$^+$ loss (total mEq)	122	244	366	172	344	516	222	444	666
Resultant plasma Na$^+$ concentration (mEq · L^{-1})	136	132	128	136	132	128	136	132	128
Fluid intake, in excess of sweat volume, to dilute plasma Na$^+$ to 120 mEq · L$^{-1\,c}$ (L)	4.2	3.2	2.2	5.9	4.5	3.0	7.6	5.8	3.9
(L · h^{-1})	0.5	0.4	0.2	0.7	0.5	0.3	0.8	0.6	0.4

[a]Running at 10 km · h^{-1} for 9 h. [b]Sweat sodium (Na$^+$) concentration decreases with physical training, heat acclimation, and low-sodium diets. [c][(initial TBW volume) × (Na$^+$ concentration after exercise) / (120 mEq · L^{-1})] = final TBW volume; fluid excess to reach 120 mEq · L^{-1} = (final TBW volume) – (initial TBW volume).

Na$^+$ = sodium; L = liters; L · 9h^{-1} = liters per nine hour; mEq = milliequivalents; TBW = total body water; mEq · L^{-1} = milliequivalents per liter.

Adapted, by permission, from S.J. Montain et al., 2001, "Hyponatremia associated with exercise: Risk factors and pathogenesis," *Exercise and Sport Science Reviews* 29(3): 113-117.

hyponatremia (118 mEq · L^{-1}) and rhabdomyolysis (serum creatine phosphokinase [CPK] of 1,545 IU · L^{-1}, increasing to 10,300 IU · L^{-1} on day 3 of hospitalization). This high CPK peak was well above what the attending physicians who recorded this patient's history thought it would be, because the normal resting range is 22 to 198 IU · L^{-1}. This led Putterman and associates (74) to recommend that serum levels of muscle enzymes be measured regularly during treatment for EH, to allow for early detection of rhabdomyolysis and employment of prophylactic measures against pigment-induced renal failure (see chapter 3).

Knochel and Schlein (73) have published a plausible explanation of the mechanism by which hyponatremia might cause rhabdomyolysis. In their theory, acute hyponatremia causes cell swelling because of the reduced extracellular fluid osmolality. However, these swollen cells return to their normal volume after 4 to 24 hours, despite continued hypoosmolality. Restoration of cell volume by the net outward movement of intracellular potassium has been demonstrated in many tissues (75); this is accompanied by an inward movement of sodium and chloride (76). The net loss of intracellular potassium may result in whole-body and muscle deficiency. During exercise, contraction of muscle normally releases potassium into the interstitial fluid surrounding muscle fibers. This causes increased blood flow to the exercising muscle by dilating adjoining arterioles (77, 78). Knochel and Schlein (73) postulated that skeletal muscle injury or death of muscle tissue (i.e., rhabdomyolysis) may occur as a consequence of relative ischemia, and they subsequently investigated groups of normal and potassium-depleted dogs. In all potassium-depleted animals, exercise-stimulated muscle blood flow and potassium release were markedly subnormal, and frank rhabdomyolysis was verified. The authors concluded that their findings may be particularly relevant to athletes and military personnel.

To discern whether rhabdomyolysis is associated with EH, researchers must examine reports of EH to identify cases that include rhabdomyolysis, verified by large increases in circulating CPK concentrations. The two disorders may be linked, as evidenced by the previous case report of the hiker (74). However, the literature shows that only a portion of all EH cases include verified rhabdomyolysis because serum enzymes were not measured. For example, among those individuals who experienced EH in table 6.2, serum CPK (or histological) verification of rhabdomyolysis was not attempted in 69% (9 out of 13) of these cases, and such events (e.g., lengthy hikes, ultramarathons) commonly induce muscle damage. Therefore, it is presently premature to link hyponatremia mechanistically to rhabdomyolysis, until clinical field studies evaluate this matter thoroughly.

TREATMENT

Although severe EH among athletes, soldiers, and laborers is usually associated with fluid overload, various fluid–electrolyte states may be associated with mild hyponatremia. Regardless of the nature of the salt and water imbalance, the severity of this disorder dictates the nature of treatment (36). Table 6.8 provides a classification scheme of EH severity, based on plasma sodium concentration and the amount of excess fluid required.

Table 6.8

Categories of Acute Exertional Hyponatremia (EH) and Onset of Symptoms During Prolonged Exercise

Category	Plasma Na+ concentration range (mEq · L⁻¹)	Approximate amount of excess fluid (L) required[a]
Mild EH	130-135[b]	0.0-0.9
Moderate EH	125-129[b]	1.2-2.4
Severe EH	<125[b]	>2.4
Approximate threshold for the onset of EH symptoms[c]	125-130[d]	0.9-2.4

[a]Using the linear regression equation in figure 6.3 and the data of Speedy et al. (11): Fluid excess = (fluid intake, L) – (sweat loss, L) – (urine volume, L). [b]Compiled from references 14, 15, 24, 25, 26. [c]Presence or absence of symptoms depends on body size, total body water, beverage composition, fluid volume, rate of fluid intake, sweat rate, and sweat sodium concentration. [d]From reference 11 and figure 6.3.

Na+ = sodium; mEq · L⁻¹ = milliequivalents per liter; L = liter.

Asymptomatic hyponatremia is not treated because these athletes do not seek medical care. Mild symptomatic hyponatremia (serum sodium of approximately 125-135 mEq Na+ · L⁻¹) may involve mild dehydration or mild **hyperhydration.** Judicious use of intravenous normal saline in these patients might be considered, but oral rehydration with salty solutions is safer, because it reduces the risk of exaggerating a preexisting fluid overload. Intravenous fluids should not be administered to a collapsed ultradistance athlete before an accurate diagnosis has been made (19). In fact, Nagara (79) recommended that athletes take fluids orally, not intravenously, if they are truly in need of rehydration, until evidence that they do not have a fluid overload is available. A physician can determine an appropriate treatment plan by observing peripheral edema and neck vein distention (both signs of extracellular volume expansion) or by observing dry mucous membranes, poor skin turgor, and orthostatic hypotension (i.e., signs of extracellular volume contraction) (25). These observations, combined with plasma sodium and osmolality values, will provide definitive guidance, as shown in figure 6.5.

Figure 6.5 and table 6.3 (presented previously) describe the three forms of acute hypotonic hyponatremia. Diuretic use may be hazardous when hypovolemic hyponatremia exists (i.e., reduced ECV) (80). In such cases, intravenous administration of normal saline is appropriate, to replace sodium and fluid deficits. If body water is normal, as in euvolemic hyponatremia, fluid restriction while the patient spontaneously excretes retained fluid is the only treatment required. Symptomatic hypervolemic hyponatremia (i.e., fluid overload with evidence of ECV expansion) is treated with a diuretic (e.g., furosemide) because it reduces ECV and causes intravenous sodium to be retained.

Severe symptomatic hyponatremia is usually associated with the retention of excess fluid, and appropriate management involves careful observations and regular monitoring of serum sodium concentrations, while awaiting spontaneous diuresis of excess fluid (81). If the patient is clinically stable and without signs or symptoms of significant cerebral or pulmonary edema, he or she should not be given intravenous fluids.

CNS manifestations of EH (e.g., mental confusion, convulsions, coma) require aggressive treatment. Seizures and coma constitute life-threatening emergencies and should be treated with hypertonic saline, to create a concentration gradient of 10 to 20 $mEq \cdot L^{-1}$, which should rapidly cause water to move out of the brain (12, 82). A rate of correction that is too rapid may result in iatrogenic complications, including the rare and serious neurologic condition central pontine myelinolysis (CPM). Most experts believe that the risk of CNS damage due to severe hyponatremia greatly outweighs the risk of developing CPM (36).

Cheng and colleagues (25) reviewed 20 years of case reports involving the management of self-induced water intoxication in psychiatric patients with polydipsia (explained previously). A total of 107 patients were treated early, before respiratory arrest or other serious complications developed. Only one of these patients, a 52-year-old woman, died, but a computed tomography scan of her head was normal, without edema. In contrast, 14 other patients were not treated or received medical care after a long delay. All of these individuals died of respiratory arrest or ventricular fibrillation. Thus, mortality among patients who received early treatment

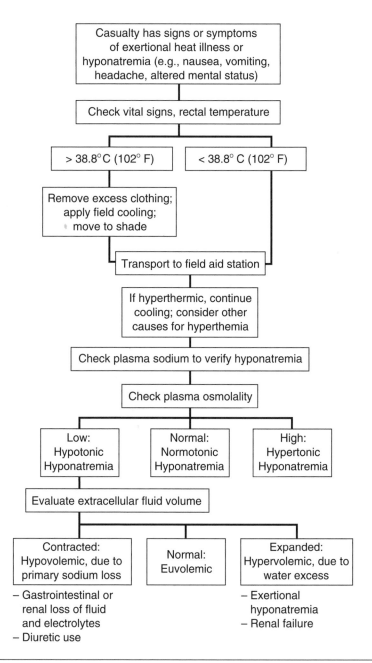

Figure 6.5 Diagnoses of specific hyponatremic states that will direct treatment. Most cases of exertional hyponatremia involve low plasma osmolality and expanded extracellular volume.

Adapted, by permission, from Flinn SD, Shere, RJ: Seizure after exercise in the heat. *Physician Sportsmedicine* 2000; 28(9): 61-67 © The McGraw-Hill Companies.

was 1%, while mortality was 100% among those not treated or treated late.

Cluitmans and Meinders (83) reviewed case reports of 163 severe, nonexertional hyponatremia patients (<121 mEq $Na^+ \cdot L^{-1}$) and analyzed their rate of development, rate of correction, treatment strategy, and outcome. Few of these cases involved exercise. They noted that recommendations regarding the rate of correction are not consistent across the international medical community. In considering this database, the authors discovered that distinguishing acute versus chronic hyponatremia was valuable, when the rate at which sodium imbalance occurred was partitioned at 0.5 mEq $\cdot L^{-1} \cdot h^{-1}$; they defined greater rates as acute and smaller rates as chronic. Rapid (acute) decline in serum sodium concentration classically develops after the intake of a large quantity of hypotonic fluids, particularly in the presence of impaired water excretion, and includes EH. Cluitmans and Meinders (83) concluded that (a) immediate treatment of acute hyponatremia is mandatory, (b) rapid correction of the serum sodium level (at least 1 mEq $\cdot L^{-1} \cdot h^{-1}$) to a mildly hyponatremic level can prevent severe morbidity or mortality, and (c) use of furosemide plus isotonic or hypertonic intravenous saline produces fewer complications than the use of saline solutions alone. These conclusions do not apply to cases of chronic hyponatremia and were supported by Cheng et al. (25), who also noted that the rapid correction of serum sodium levels during the initial 12 hours should be followed by a slower correction via fluid restriction.

The aforementioned review of the literature is not intended to suggest that all authorities agree on the optimal management of acute symptomatic hyponatremia. For example, Berl (84) stated that correction of this condition carries no harmful consequences, regardless of the treatment regime. Sterns (85) recommended correction at 1 to 2 mEq $\cdot L^{-1} \cdot h^{-1}$. Ayus, Krothapalli, and Arieff (86) proposed a sodium correction rate of 1 mEq $\cdot L^{-1} \cdot h^{-1}$, until the patient becomes asymptomatic or until a serum level of 120 mEq $Na^+ \cdot L^{-1}$ is attained, but not to exceed an increase of 25 mEq during the first 48 hours of therapy. The latter guideline arose from their belief that the absolute change over 48 hours is more important than the rate of change. Further, one author suggested that the American College of Physicians should generate a position paper on the treatment of this disorder (87). Today, this still appears to be a sound recommendation. A thorough comparison of treatments, sodium correction rates, and patient outcomes appears in the review published by Arieff (28).

Davis et al. (14) reported that hypertonic saline (mean: 360 ml of 3% NaCl) was tolerated well, without neurologic sequelae, in five female marathon runners, ages 24 to 55, with acute, severe EH (serum sodium <125 meq · L^{-1}) who were treated at a medical center. The serum sodium correction rate (3.4 meq · L^{-1}· h^{-1}) was considerably faster than the rate administered one year earlier (1.8 meq · L^{-1}· h^{-1}) to patients after the previous year's marathon. This faster correction rate resulted in fewer seizures (i.e., none) and fewer inpatient admissions (i.e., none). Similar success with hypertonic saline therapy has been reported for a hiker with both cerebral and pulmonary edema (5), a military recruit (9), an ultradistance triathlete (23), a healthy young male (15), and hospitalized patients (28).

PREVENTING EXERTIONAL HYPONATREMIA

Because the vast majority of EH cases today result from overhydration, virtually all of these are preventable. Prevention and risk reduction can be accomplished by modifying sodium intake and moderating fluid intake. Psychological factors and attitudes toward water consumption are also worthy of consideration.

Dietary Sodium

People participating in prolonged endurance events (>4 h), should ingest electrolytes (e.g., sodium, chloride, potassium, magnesium) via fluids and food during and after exercise. The American College of Sports Medicine (ACSM) position stand "Heat and Cold Illnesses During Distance Running" acknowledges the potential for severe dilutional hyponatremia to cause illness, seizures, and coma (7). This supports the rationale for including sodium in fluid replacement beverages, although the effect likely will be small.

Presenting runners with guidelines regarding sodium intake also may reduce the risk of EH (88). For example, brief rest periods should be taken during ultraendurance events to allow for consumption of sodium-rich beverages and foods (e.g., low-fat soup or stew with crackers). One research team even developed a sodium-enhanced water solution (0.06% NaCl). This level provided the highest concentration of salt that was palatable in water, without adding glucose or flavoring (9).

Two opposing viewpoints have emerged regarding the best dietary plan to follow during the 3 to 5 days before prolonged exercise—one recommending against aggressive rehydration with water and one recommending against overconsumption of sodium during the days prior to an event that involves consumption of a large quantity of water. First, one case report observed a "low normal" plasma sodium level (134 meq · L^{-1}) on the day before symptomatic hyponatremia occurred. The authors of this report (15) observed that aggressive prehydration with pure water predisposed an otherwise healthy human to EH. Second, an animal model of hyponatremia clarified the importance of sodium in the development of this disorder (89). Rats received isotonic saline to expand their extracellular fluid volume, but no oral food or liquid, with and without exogenous administration of AVP. Hyponatremia did not develop if the rats consumed a low-salt diet for 3 days prior to the experiment. Additional human research is required to clarify the best preventive dietary strategy and to evaluate an optimal hormonal status (e.g., aldosterone, AVP) before ultraendurance exercise.

Fluid Consumption Guidelines

The only certain method of preventing EH is to avoid drinking too much pure water or dilute fluid (14, 24, 36, 41). This approach has been recognized by numerous authorities in the areas of fluid–electrolyte balance and field emergency care. For example, the ACSM position stand "Exercise and Fluid Replacement" (90; appendix A) suggests that fluid be consumed at a rate that matches sweat losses (16). However, this recommendation was written as general guidance for all sports and may not specifically apply to ultraendurance events. Montain, Sawka, and Wenger (16) cautioned that competitive and recreational athletes, as well as laborers, should be taught that excessive fluid intake can be harmful. Garigan and Ristedt (4) warned that excessive intake of fluids may lead to hyponatremia and potentially fatal cerebral and pulmonary edema. Speedy et al. (26) stated that sensibly limiting the intake of appropriate fluids is the most important intervention for this condition. As mentioned previously and in table 6.4, the rate of fluid consumption may be as important, or more important, than the total volume, in the development of EH.

Evidence suggests that intense determination to drink a large volume of dilute fluid may play a role in the development of EH. The author and colleagues (7) published an example of this behavior in 1996. A 21-year-old male unexpectedly experienced EH while serving as a participant in a research study. He began the initial day of exercise in an environmental chamber with a "low normal" initial plasma sodium level (134 mEq · L^{-1}) because he had consumed so much water on the previous day (see table 6.5). The subject believed that consuming water excessively provided immunity from heat illnesses. During the first day of testing (41°C, 105.8°F; 30 min exercise · d^{-1}, 8 h · d^{-1}), the test subject voluntarily consumed 10.3 liters of pure water in 7 hours, to the surprise of the investigators, resulting in a plasma sodium concentration of 122 mEq · L^{-1} and clinical symptoms of EH. At least two other research teams have reported similar observations. Garigan and Ristedt (4) observed that "dogged single-mindedness" caused a healthy 18-year-old military recruit to continue drinking, even as his symptoms became worse. And, in eastern Nepal, a 28-year-old athletic woman developed classic signs and symptoms of EH (with a serum Na^+ concentration of 122 mEq · L^{-1}) during a low-altitude trek through the Himalayan mountain range. The authors of this case report (91) noted that the trekker strongly believed in drinking plenty of fluids (5-6 L) anytime she worked out, and she believed that her symptoms could be overcome by drinking even more water. She ingested about 10 liters of water that day. Noakes and colleagues (20) observed the drinking behavior of several runners who previously had experienced hyponatremia. These researchers found that altering fluid intake purposefully, via instructions, prevented the runners from a recurrence of this disorder. This suggests that education can successfully reduce the incidence of EH. These four reports illustrate the importance of educating active individuals about the need to consume an *adequate*, but not *excessive*, volume of fluids. This clearly is a key to preventing EH.

SUMMARY

Mild exertional hyponatremia is defined as a serum sodium concentration between 130 and 135 mEq · L^{-1}. The prominent signs and symptoms include disorientation, depression, nausea, vomiting, muscular twitching, and grand mal seizure. Severe cases (serum sodium concentration <125 mEq · L^{-1}) may involve coma, pulmonary or cerebral edema, or respiratory arrest. Only this disorder and heatstroke are potentially fatal to otherwise healthy athletes, laborers, and military personnel. Despite these dangers, few athletes realize that excessive fluid consumption may cause illness, hospitalization, cognitive impairment, or occupational disability.

Reports of EH most often arise from military training, long hikes, distance running events (≥42.2 km, ≥26.2 miles), and triathlons (7-17 h). Virtually all individuals who experience EH attempt to drink as much fluid as possible during and after exercise. During prolonged exercise, a rate of fluid consumption of about 0.7 L · h^{-1} represents an approximate threshold in the development of EH signs and symptoms. Body size, total body water, beverage composition, sweat rate, and sweat sodium concentration are important etiological factors. The simplest way to reduce the risk of EH is to ensure that fluid is consumed at a rate that equals or is less than the sweat rate. Postexercise consumption of sodium-rich beverages and foods (e.g., low-fat soup or stew with crackers) also reduces EH risk.

Chapter

Minor Heat Illnesses

Robert W. Kenefick, PhD, FACSM
Melissa P. Hazzard, MS
Lawrence E. Armstrong, PhD, FACSM

Miliaria Rubra in a Young Soldier: Living and Training in a Hot–Wet Environment	Case Report 7.1

R.G. was a 26-year-old Caucasian male (188 cm, 73 in; 92.7 kg, 205 lb) with a history of intermittent skin rash, since the age of 14 years, caused by 5 to 14 days of repeated heat exposure. He had no previous history of exertional heatstroke or heat cramps. A soldier, R.G. was admitted four times in 3 years with miliaria rubra (i.e., prickly heat) on the abdomen, back, shoulders, chest, and lateral surface of the arms and elbows. These are pressure areas for a wet T-shirt. Treatment consisted of lotions to control mild desquamation and restriction of exposure to heat, humidity, and sweating. R.G. was stationed in the southern United States at a military training facility. His daily schedule included physical exercise and recreational sports (primarily softball and American flag football) in a hot–wet environment. R.G. rarely took time to shower during field maneuvers and thus spent many hours each day in a sweat-soaked uniform. After his admission to the local medical clinic, R.G. was referred to a research laboratory for an 8-day evaluation of sweat gland function, heat acclimation, and tests of thermoregulation.

During days 1 and 2 of laboratory experiments, miliaria rubra was visible on his back, upper arms, shoulders, and chest. R.G. occasionally reported a mild stinging sensation, but this did not include itching. Despite daily standing and exercise in the heat (40°C, 104°F; 48% relative humidity), plus heavy sweating (1.4-2.6 L · h⁻¹), the miliaria rubra was not visible on days 3 to 8. A skin biopsy over his right scapula on day 8 was negative for miliaria rubra but was positive for chronic folliculitis (i.e., inflammation of hair follicules). Miliaria rubra may be a precursor to folliculitis. The heat-acclimation program induced the expected adaptations of heart rate, rectal temperature, sweat rate, and sweat electrolytes. Fluid–electrolyte balance and urine indices of hydration were normal. Sweat sodium decreased from 57 to 23 mEq $Na^+ \cdot L^{-1}$ on days 2 and 7, respectively. During this period, whole-body sweat rate increased from 1.35 to 2.16 liters per hour. The number of heat-activated sweat glands (per square cm) did not change during this heat-acclimation protocol. Eight hematological measures and 15 blood biochemical analyses were unremarkable.

Therefore, the following conclusions were drawn. First, miliaria rubra subsided during 8 days of a heat-acclimation program that included both heavy sweating and 90 minutes of exercise in a hot environment each day. Daily showers likely were instrumental in this outcome. Second, despite the fact that miliaria rubra impairs heat tolerance, R.G. exhibited normal heat-acclimation adaptations. Third, other investigations have shown that both saltwater bathing and a high sweat sodium chloride (NaCl) concentration are predisposing factors for miliaria rubra. Because R.G. exhibited a decreased sweat NaCl concentration during heat acclimation, he was less likely to maintain miliaria rubra. Finally, the first and third items suggested that R.G. could prevent subsequent problems by daily skin cleansing and participating in a heat-acclimation program prior to prolonged exposure to a hot–humid environment. (This report was revised and condensed from reference 16.)

Minor heat illnesses—for example, those involving fatigue, anhidrosis, edema, dermatologic conditions in the miliaria family, and sunburn—are considered benign conditions, in comparison to heat exhaustion and heatstroke. However, preventing and treating these disorders can limit the pain and discomfort associated with them and keep them from developing into serious conditions. Although these minor heat illnesses are more common in the summer months, they may occur in moderate conditions where laborers work in severe heat (e.g., factories) and are related to physical exertion. The purpose of this chapter is to define and describe the causes and symptoms of each minor heat illness and to make recommendations regarding treatment and prevention.

TRANSIENT HEAT FATIGUE

Transient heat fatigue (i.e., mild heat fatigue, acute heat neurasthenia) is a condition attributable to short-term exposure to a hot environment. It typically occurs in the summer months or when the weather is hot, but it is also commonly reported by individuals who work in poorly ventilated spaces (e.g., factories, warships) or with equipment that emits excessive heat.

Transient heat fatigue refers to the temporary state of discomfort and mental or psychological strain arising from heat exposure. Symptoms include a decline in performance, particularly in skilled physical work, mental tasks, and those requiring concentration (1, 2). The frequency of workplace accidents appears to be higher in hot environments compared to moderate environments because of the mental and physiological fatigue associated with heat-related illnesses such as transient heat fatigue. The declines in mental alertness and physical performance and increases in body temperature and physical discomfort promote irritability, anger, and other emotional states. Such changes in emotional state and mental acuity sometimes cause individuals working with dangerous equipment or in dangerous situations to become less aware of safety procedures and safety hazards (3).

The main cause of transient heat fatigue is discomfort from the heat, and this discomfort is less than that which would result in other heat illnesses. Individuals unacclimated to the heat are particularly susceptible to mental or physiological strain arising from prolonged heat exposure and can suffer varying degrees of decline in task performance, coordination, alertness, and attention. No specific treatment outside of rest and rehydration is necessary unless other signs of heat illness are present.

Heat disorders such as transient heat fatigue are more likely to occur among those who were not given time to adjust to working in the heat or those who have been away from a hot environment for an extended period. The severity of transient heat fatigue has been reported to be lessened by a period of gradual adjustment to the hot environment (heat acclimation) and physical training (2). Minimizing exposure to heat in work areas can reduce the risk of transient heat fatigue. If work in hot places cannot be avoided, then the strain can be minimized if the heat acclimation is distributed over at least 5 days and if hydration is maintained during the acclimation process.

In work situations (factories, warships, and so on), individuals may be able to acclimate themselves to the heat by rotating in and out of the heat in shortened work shifts, increasing work shift time over a 5 to 7 day period until they are working a full shift, or exposing individuals (at rest) to heat several days prior to working in the heat. It should be noted that because these methods involve limited exposure and work in the heat, they may require a longer time frame to achieve acclimation.

ANHIDROTIC HEAT EXHAUSTION

All three types of heat exhaustion—water deficiency heat exhaustion, salt deficiency heat exhaustion, and anhidrotic heat exhaustion—can be debilitating. Prior to and during World War II, numerous cases of soldiers suffering from a syndrome termed *tropical anhidrotic asthenia* or *heat exhaustion II* were reported by medical personnel in tropical and desert regions (2). The term *anhidrotic heat exhaustion* is more commonly used today. Anhidrotic heat exhaustion is a rare malfunction of the sweat glands that occurs in people who have been in a hot climate for several months. Early reports of this condition (2) describe a syndrome involving the failure of the normal sweat mechanism in a group of soldiers affected by extreme heat. Although some of the symptoms of this condition are similar to heatstroke, the conditions exhibit noteworthy differences.

Anhidrotic implies a complete absence of all sweating, which is also characteristic of heatstroke. In both classical heatstroke and anhidrotic heat exhaustion, the skin is dry and sweating ceases or markedly declines. However, individuals with heat exhaustion usually have headache, generalized extreme weakness, mental exhaustion, and vertigo. Although the cessation of all sweating never truly occurs in anhidrotic heat exhaustion, the term is used to imply a reduction in sweating, usually with complete absence of sweating on the trunk, but notably with sweating still occurring on the face. As anhidrotic heat exhaustion is almost always preceded by prickly heat (miliaria rubra), its cause is likely related to a decrease in sweat gland activity resulting from the miliaria rubra. These overall symptoms are typically brought about by physical exertion in the heat. Other symptoms include **polyuria,** increased frequency of urination, and dry and unusual skin texture.

Anhidrotic heat exhaustion typically begins with the presentation of miliaria rubra. The rash is usually severe, extensive, and recurrent and most often disappears 3 to 4 weeks prior to anhidrosis. The subsequent onset of anhidrotic heat exhaustion is gradual. Most often individuals notice that exercise in the heat produces an unusual degree of exhaustion, dyspnea, and headache. In addition, some individuals report a hot, congested, or tight prickling feeling in the skin. In the early stages of the condition, sweating decreases. As the condition progresses, the symptoms reemerge more readily and severely with decreasing amounts of exertion. During flare-ups of anhidrotic heat exhaustion, the face is flushed and sweating, and the skin of the trunk exhibits mammillaria, while the whole body (except for the face) is pale and dry (1).

Evidence suggests that anhidrotic heat exhaustion is a result of peripheral failure of sweat secretion and not a failure of a central regulatory mechanism. Sweat chloride concentrations and pH are elevated in many cases of the syndrome (4). The onset of symptoms principally is dependent on the person's level of exertion and on the air temperature. The syndrome has been reported in both humid and desert climates. The acute symptoms of this condition are usually rapidly relieved in about 30 minutes to 2 hours by rest and cooling. Anhidrotic heat exhaustion can last from 2 weeks to 3 months, with an average of about 6 weeks. Management of this condition requires removal from excessive heat until sweat function returns to normal. Increasing dietary sodium chloride intake has not been shown to be effective in the treatment of this condition (2). Recovery should be judged by the ability of individuals to withstand exposure to a hot environment without symptoms.

HEAT EDEMA

Heat edema is one of the most minor forms of heat-related illness, but it is quite common in individuals with poor circulation, such as the elderly, and in individuals not acclimated to heat. As fluid shifts occur between the intracellular and extracellular compartments of body tissues, the abnormal accumulation of fluid within the interstitium, also known as edema or swelling, may occur (chapter 2). Individuals experiencing swelling in the legs and arms while exposed to hot environments may be experiencing heat edema (5, 6).

Although the pathophysiology of heat edema is not known, this swelling of the extremities may be caused by a combination of factors, including hydrostatic pressure, vasodilation of the blood vessels, and orthostatic pooling of blood occurring with prolonged periods of standing and sitting (6). In addition, when people are exposed to high heat and humidity, they lose a significant amount of fluid via sweating. To maintain fluid and sodium levels during such conditions, aldosterone is released by the adrenal glands of the kidneys. Although aldosterone limits muscle cramping via the retention and maintenance of sodium levels, the result is the accumulation of fluid, particularly in areas where gravitational forces play a role (4).

Improving tolerance to hot–humid environments, primarily through heat acclimation, can combat the susceptibility to heat edema. In addition, limiting standing or sitting time and elevating the feet while in the heat can help limit the impact of gravitational forces. During exercise, skeletal muscle contractions are able to relieve venous pressure by maintaining blood flow to the heart, but during long periods of standing, fluid typically accumulates in the legs.

MILIARIA

The family of skin disorders named miliaria includes a broad spectrum of heat illnesses. First recognized in soldiers during World War II, the term *miliaria* may be applied to any skin condition resulting from the retention of sweat within the skin (7, 8). Although infants are most commonly diagnosed with this skin disorder, likely because of the immaturity of their eccrine sweat glands (9), individuals constantly exposed to hot and humid environmental conditions are also highly susceptible. Miliaria is notably seasonal in occurrence, commonly diagnosed during the spring and summer months (5, 7, 8, 10-12).

While the pathogenesis of miliaria is still unclear (13), evidence shows that the problem is not necessarily the inability to produce sweat but rather the inability for this sweat to flow freely to the surface of the skin. In children and adults, this inability to dissipate sweat is likely caused by an injury to the epidermis and the development of keratin (i.e., a type of tough protein) plugs of the sweat ducts, resulting in the occlusion of the terminal end of the ducts and the retention of sweat within the skin (8). When occlusions occur within different levels of the ducts, sweat begins to accumulate within the skin layers, and, after approximately 48 hours, miliaria and consequently anhidrosis occur (7). In addition, as sweat is retained within the skin, layers become overhydrated, creating a moist environment and ideal conditions for bacterial colonization. The inflammation accompanying miliaria is in part caused by the increase in the presence of bacteria and the leakage of retained sweat into the surrounding tissue (12). Because miliaria reduces sweat dissipation, the body is less able to maintain body temperature homeostasis and to tolerate hot–humid ambient conditions (5, 7, 8, 11, 12, 14). Thus, miliaria may lead to severe hyperthermia or heatstroke during physical exertion or when exposed to extreme environmental heat at rest (5).

The four classifications of miliaria are miliaria rubra, miliaria pustulosa, miliaria profunda, and miliaria crystallina. Each classification may be identified by the layer at which the occlusion has occurred and by the characteristics of the vesicles, pustules, and papules erupting on the skin surface. In general, miliaria is characterized by the presence of numerous nonfollicular lesions approximately 1 to 3 millimeters in size, typically confined to the torso and upper extremities (8, 12). The severity of the outbreak of miliaria depends on the percentage of surface area affected, the abundance of sweat produced at the site, and the extent of heat stress the body is being exposed to at the time, whether from environmental heat or from heat production during physical activity (15).

Miliaria Rubra

Miliaria rubra, commonly called "heat rash" or "prickly heat," is the most minor of the miliarias (5). Miliaria rubra may be identified as a fine, red, papular rash typically confined to the torso, neck, and skinfolds of the body (7, 8, 15, 16). The severity of the rash varies with fluctuations in body temperature, rapidly appearing as the body begins to get hot and disappearing as the body cools. The site of the miliaria rubra outbreak may be inflamed, and severe itching and burning sensations may occur at the site with the onset of sweating (7, 8, 11, 15, 16).

Characterized as intermediate in the depth of occlusion, this form of miliaria develops in response to duct obstruction at the **intraepidermal layer,** resulting in the accumulation of sweat within the deeper layers of the epidermis (10). This obstruction is caused by the formation of a periodic acid plug combined with the presence of inflammatory cells passing into the epidermis (17). If heat exposure is minimized, this plug disappears as the epidermal layer renews itself. The epidermis, or outer layer of the skin, is composed of a thin sheet of skin cells built up from many layers of flat cells produced continuously in the deepest region of the epidermis. The epidermal layer renews itself as the outer most skin cells slough off and new, mature skin cells push up from the deeper region of the epidermis. This process may take as long as 3 weeks, at which time the sweating mechanism and heat tolerance is restored (16).

Interestingly, the incidence of miliaria rubra depends on the duration of a group's residence in a hot–humid climate. Soldiers who were transferred to Southeast Asia experienced an increased incidence from 0 to 4 months (i.e., the peak); a reduced but level incidence from 12 to 15 months; and a decreasing incidence from 15 to 21 months (18). A significant reduction in physical performance has been found after a single episode of miliaria rubra and may continue for approximately 14 days after the condition is remedied (15).

Miliaria Pustulosa

Miliaria pustulosa is diagnosed when the small vesicles of miliaria rubra become pustular in nature, with white, and sometimes yellow, purulent material contained within the vesicles (8, 11). Typically occurring at a site of previous skin disease or inflammation, such as that of miliaria rubra, the superficial vesicles of miliaria pustulosa may contain bacteria, specifically either pathogenic or nonpathogenic staphylococci (8, 11). Anhidrosis typically does not accompany miliaria pustulosa because this form of miliaria affects only a small percentage of sweat glands (8).

Miliaria Profunda

Miliaria profunda is the deepest form of miliaria and typically is reported after repeated outbreaks of miliaria rubra. Identified as firm, pale, or flesh-colored noninflammatory papules approximately 1 to 3 millimeters in size, miliaria profunda is typically limited to the trunk and extremities. These papules may become enlarged during physical activity or with exposure to extreme hot and humid conditions, and are stimulated by the onset of sweating (8, 11).

Because of a duct obstruction at the **dermoepidermal junction,** this noninflammatory form of miliaria develops in response to the retention of sweat within the dermis portion of the skin (8, 10-12). As **keratin plugs** form within the ducts, sweat is retained within the dermis, and the sweat ducts within the dermoepidermal junction begin to dilate and eventually rupture. This leads to some local swelling at the site of the rupture and the formation of the firm papules of miliaria profunda. Only when the epidermis is renewed is sweating restored at this site (11).

Individuals diagnosed with partial to complete anhidrosis, which occurs with miliaria profunda, are at high risk for hyperthermia because the sweat glands become inactive (10, 12). Severe cases of miliaria profunda may lead to symptoms of tropical anhidrotic asthenia, including weakness and fatigue, syncope, nausea, and dyspnea (11).

Miliaria Crystallina

Miliaria crystallina is the most superficial form of miliaria (10) and is rarely diagnosed (12). Newborns are particularly susceptible to miliaria crystallina because of the immaturity of the acrosyringium and exposure to the moist intrauterine environment (9). Miliaria crystallina, however, has also been associated with severe sunburns (9, 10).

Miliaria crystallina may be identified by the superficial eruption of clear, noninflammatory vesicles approximately 1 to 3 millimeters in size (8, 9, 10, 12). In this syndrome, the occlusion occurs within the ducts of the **intracorneal** or **subcorneal layers** of the skin, with sweat trapped within the outer layer of dead skin (8, 9). Some vesicles may appear white if erupting under a thick layer of the **stratum corneum.** The lesions of miliaria crystallina commonly rupture after the cessation of sweating (9).

Mammillaria is similar in nature to miliaria rubra; it is associated with anhidrotic heat exhaustion and becomes more severe as exposure to hot and humid conditions increases. Typically found after an outbreak of miliaria rubra (4), mammillaria may be identified by the eruption of pale, circular elevations on the skin approximately 1 millimeter in diameter. Occurring particularly between the neck and the waist, these "cobblestone" elevations appear and disappear according to the amount of heat exposure. Over time, the diagnosis of mammillaria becomes more difficult, because even as the condition persists, the small elevations in the skin become less sensitive to heat exposure and begin to flatten out (4).

Mammillaria may persist for varying lengths of time. In fact, cases have been reported for as long as four months, and full recovery of the skin to normal can be a slow process. As both mammillaria and anhidrotic heat exhaustion seem to develop after severe bouts of miliaria rubra, prevention and treatment of these conditions follow that of other forms of miliaria.

Treatment of Miliaria

The primary goal of treatment for miliaria is to ensure that normal sweat production resumes. The treatment for miliaria is to remove the source of sweating, whether the metabolic production of heat in response to exercise or in response to heat gain from the environment (8). A break from exercise and rest in a well-ventilated, air-conditioned room reduces the need to thermoregulate via sweating (12). Affected individuals should avoid or remove factors that lead to duct obstruction, such as exposure to high humidity and sodium on the skin. Showering regularly helps remove both salt and bacteria from the skin (12, 19); however, excessive handling with soaps and detergents may be harmful, as the lesions of miliaria may rupture. Superficial forms of miliaria, such as miliaria crystallina, are easily irritated by constrictive clothing and excessive palpation (8, 11). In most cases, an individual diagnosed with miliaria will see improvements after only 2 to 3 days of treatment. The symptoms of this skin disorder however, may not be fully resolved for at least 2 weeks, until the keratin plug within the ducts is shed and the epidermis renews itself (11, 13).

SUNBURN

When exposed to excessive ultraviolet radiation, the epidermal and dermal layers of skin exhibit injury or phototoxicity that is commonly known as sunburn. Ultraviolet-B (UV-B), with a wavelength of 280 to 320 nanometers (nm), is the type of ultraviolet radiation that is responsible for sunburn. Approximately 90% of the energy in UV-B waves is absorbed by the epidermis, and the remainder is absorbed by the dermis. This results in a cutaneous reaction that presents as pain, erythema (i.e., redness), and edema, beginning several hours after exposure, peaking at 14 to 20 hours, and resolving in 24 to 48 hours (20). These three clinical signs represent an inflammatory reaction that can be reduced by using nonsteroidal anti-inflammatory drugs (NSAIDs). Exposure to UV-B also decreases immunity in cutaneous cells and alters systemic function of the immune system (21).

A recent study (22) demonstrated that the incidence of sunburn is high in certain demographic groups. Nationwide telephone interviews of 156,354 Americans showed that 32% of all adults experienced sunburn during the previous 12-month period. Among these, the highest incidence occurred in the 18- to 29-year-old age group. Ethnicity also played a role in incidence. White, non-Hispanic males (44%) and females (35%) reported the greatest number of sunburns, whereas Black, non-Hispanic males and females reported the smallest incidence (5% for both groups).

Human skin becomes sensitized to UV-B radiation if certain substances are ingested or applied to the skin. Warning labels on many widely prescribed medications describe this enhanced sensitivity to sunlight. Table 7.1 provides a list of selected substances that increase photosensitivity and photoallergic reactions. Individuals should limit the use of such agents when prolonged exposure to UV-B radiation is anticipated. Further, individuals should limit unnecessary exposure to direct sunlight, because of the increased risk of skin cancer (i.e., melanoma) (20).

As with miliaria rubra, sunburn can impair eccrine sweat gland function (20). Because sweat glands lie in the dermal layer of skin, the sweat ducts pass into the epidermis, and the cells lining the ducts contain no protective pigment, sweat glands are vulnerable to penetrating UV-B radiation. Following this line of reasoning, Pandolf and colleagues (20) evaluated the effects of sunburn on human temperature regulation during cycling exercise (50 min; 49°C, 120.2°F; 20% relative humidity). Their observations indicated that sunburned skin, compared to normal skin, exhibited a decreased ability to generate sweat. This resulted from a reduced glandular sensitivity to an increase in core body temperature and from a reduced secretory capacity in sweat glands. Sunburn also enhanced subjective distress, measured with ratings of perceived exertion and thermal sensation.

Medical treatment for sunburn includes analgesics and cool compresses for first- and second- degree sunburns. Aspirin should reduce the severity of inflammation and provide adequate analgesia, when administered before or immediately after exposure to UV-B radiation. More extensive sunburns may be treated with systemic steroids (23). To prevent sunburn, individuals should reduce expo-

Table 7.1

Selected Substances That Increase Skin Sensitivity to UV-B Radiation and Increase the Risk of Sunburn or Cause Allergic Reactions When Exposed to Sunlight

Category	Substance	Examples
Topical or local	Acne medication	Vitamin A
	Cosmetics	Indelible lipsticks, perfumes, and aftershave lotions containing essential oils
	Plants	Furocoumarins
	Deodorants in soaps	Hexachlorophene, dichlorophene
	Sunscreens	Para-aminobenzoic acid (PABA), diagalloyl trioleate
	Pigments and dyes	Yellow cadmium sulfide, proflavine, eosin
Systemic	Antihistamines	Benadryl
	Dyes	Methylene blue, fluorescein, eosin
	Food additives	Saccharin, cyclamates
	Sulfa drugs and analogues	Sulfonamides, some diuretics
	Antibiotics	Tetracycline, doxycycline
	Miscellaneous	Birth control drugs, furosemide (Lasix)

Numerous other substances exist that have similar properties.
Adapted with permission from Amundson LH: Managing Skin Problems in Athletes. In Mellion MB, Walsh WM , Shelton GL (eds) The Team Physician's Handbook, Hanley & Belfus, 1990, 236-250.

sure to the sun between 10:00 A.M. and 2:00 P.M., wear protective clothing and a hat, and apply a sun-block lotion (sun protection factor >15) before exposure. Some scientists have questioned whether skin lotions might reduce heat dissipation. Presently, experimental findings do not provide a clear answer; both no effect (24) and a small decrease in heat dissipation (25) have been reported.

Finding shade during competition can help prevent sunburn and other minor heat illnesses.

SUMMARY

Table 7.2 describes the signs, symptoms, and treatment associated with the minor heat illnesses. Early recognition of these illnesses is important because several decrease the ability to thermoregulate in the heat and lead to more severe illnesses. The incidence of most of these conditions can be reduced by gradually exposing individuals to hot environments for 10 to 14 days (i.e., heat acclimatization). Treatment of these disorders often involves restricting heat exposure until the ability to thermoregulate returns to normal. In many cases, individuals who have suffered and recovered from these illnesses may be more susceptible to recurrence. A physician or a dermatologist will offer treatment options and provide ways to reduce the risk of further complications.

Table 7.2

Signs, Symptoms, and Treatment of the Minor Heat Illnesses

Condition	Signs and symptoms	Treatment
Transient heat fatigue[a]	Changes in emotional state associated with an increase in body temperature and physical discomfort; anger and irritability; decrease in mental acuity, attention to detail, coordination and physical performance (1-3).	Immediate rest in a cool area; provide rehydration. Heat acclimation (10-14 days) will enhance heat tolerance (2).
Anhidrotic heat exhaustion	Dry, pale skin on the trunk due to a reduction in sweating; flushed face; headache; generalized weakness; mental exhaustion; vertigo. Often preceded by a bout of miliaria rubra (1).	Rest and remove the individual from the hot environment, to return sweat function to normal (2).
Heat edema	Abnormal accumulation of fluid in the extremities, due to heat exposure. Common in non-acclimatized and elderly individuals (5, 6).	Limit time of standing or sitting in a hot environment. To limit the effect of gravity on fluid shifts, assume a supine position and elevate the extremities. Heat acclimation (10-14 days) will limit susceptibility (5, 6, 26).
Miliaria	Anhidrosis, inflammation, possible hyperthermia due to dry skin (7, 12, 14).	Avoid hot-humid air and sweat-soaked or restrictive clothing; remain in well-ventilated, air-conditioned rooms; shower the skin at least once per day (8, 11-13, 19).
Sunburn	Erythema (i.e., redness), pain, edema (19, 23).	1st- and 2nd-degree burns: cold compress and analgesics. 3rd-degree burns: administer systemic steroids (23).

[a] Also known as mild heat fatigue or acute heat neurasthenia.

149

Chapter

Predisposing Factors
for Exertional Heat Illnesses

Lawrence E. Armstrong, PhD, FACSM
Douglas J. Casa, PhD, ATC, FACSM

The causes, organ systems, and symptoms associated with each exertional heat illness are unique. Thus, it should not be surprising that the predisposing factors, that is, those things that increase a person's susceptibility to an illness, are distinctive. This chapter describes the numerous predisposing factors for heat illnesses.

If an athlete, laborer, or soldier is *predisposed* to exertional heat illnesses, he or she has a heightened potential to develop that disorder in the presence of specific environmental stimuli or stressors (i.e., those factors that disrupt **homeostasis)**. As noted in chapter 1, these environmental stressors include high ambient temperature, humidity, and solar radiation and large fluid–electrolyte losses. What factors, then, predispose a person to heat illness?

EXERTIONAL HEATSTROKE

Lengthy lists of purported predisposing factors for heatstroke have been published by numerous authors (1-8), even though lack of knowledge, poor judgment, insufficient medical staff, or even neglect may have been involved in assessing the causes (9). To avoid unsubstantiated claims regarding predisposing factors, the following list presents environmental and host factors that are supported by accepted biophysical or physiological principles or published clinical observations.

- Virtually all scientific and clinical reports of exertional heatstroke involve males. One or more of the following hypotheses may explain this (1): (a) males are predisposed to exertional heatstroke because of inherent hormonal, physiological, or morphological differences; (b) males are placed in situations that result in exertional heatstroke more often than females are; (c) males readily push themselves to the point of severe hyperthermia or collapse more so than females do; and (d) scientific and clinical research historically has reported findings involving male test subjects rather than female test subjects.

- High ambient temperature, humidity, and solar radiation coupled with low wind speed reduces the body's capacity to dissipate heat to the environment. Therefore, the risk of heatstroke increases as environmental stress increases (3). Various indices, such as the discomfort index (3) and wet bulb globe temperature (10), utilize measurements of air temperature, humidity, and solar radiation to evaluate environmental stress in terms of physiological strain. But heatstroke may also occur on relatively cool days (1, 3). Interestingly, a prospective study of Marine recruits reported that the incidence of heat illness was greatest when soldiers had been exposed to hot weather on the previous day, not necessarily at the time of the episode. The investigators (11) concluded that severe environmental heat stress had some influence that lasted overnight. Hubbard and colleagues (9) have interpreted this phenomenon to be a predisposing factor lasting for at least one day.

- Enclosed vehicles, certain work sites (e.g., oil tankers, deep mines, nuclear reactors), and playing fields (especially those with synthetic surfaces) are characterized by high ambient temperatures, elevated humidity, and low wind speed. Labor or exercise at these sites predisposes people to exertional heat illnesses.

- During exercise, individuals who have experienced exertional heatstroke before exhibit a greater reliance on carbohydrate (i.e., anaerobic) metabolism than do control subjects (1). However, the effects of metabolism or muscle fiber type are inconclusive (12).

- The cardiovascular systems of older adults are less able to increase cardiac output during exercise in the heat, which normally would increase blood flow to the skin, helping to cool the body. This is especially true for individuals with congestive, ischemic, arteriosclerotic, or hypertensive heart disease. Obesity and a reduced sweating response also are predisposing factors in older adults (2). During the 1982 religious pilgrimage to Mecca (figure 8.1), age was an important predictor of heatstroke incidence (13). In the South African mines, Strydom (14) found a marked increase in exertional heatstroke with increasing age of laborers. In his sample, men over 40 years of age represented less than 10% of the total work force but accounted for 50% of the fatal and 25% of the nonfatal cases of heatstroke. The incidence of cases was at least 10 times greater for men over 40 years of age than for men under 25 years of age.

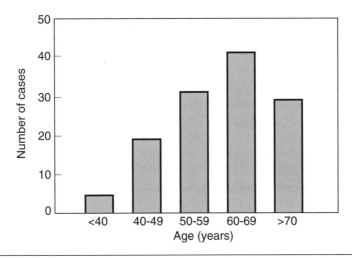

Figure 8.1 Age distribution of 125 exertional heatstroke cases during the 1982 religious pilgrimage to Mecca.

Reprinted, by permission, from M.I. Al-Khawashki et al., 1983, Clinical presentation of 172 heat stroke cases seen at Mina and Arafat–September 1982. In *Heatstroke and temperature regulation,* edited by M. Khogali, et al. (Australia, Academic Press Australia) 99-108.

- Soldiers and athletes often perform physical tasks in hot environments that exceed their fitness and acclimatization levels. This is especially true when healthy individuals are placed under the peer pressure and discipline of work, training, or competition. In other circumstances, they would have rested when tired, consumed liquid when thirsty, and remained at home when ill (5).

- Strenuous exercise increases metabolic heat production by more than ten-fold. If individuals are not capable of cooling via evaporation, convection, and radiation, they could experience a whole-body temperature rise of at least 5°C (9°F) per hour.

- A previous experience with heatstroke has been cited as a predisposing factor for a second episode (15-17), but this is not always true. For example, one evaluation of 10 prior heatstroke patients (1) demonstrated that transient heat intolerance existed in a few cases, but no evidence of permanent intolerance was observed (see case report 8.1).

- Medications may negatively affect heat dissipation (18) by reducing sweat production (e.g., antihistamines, anticholinergics), altering skin blood flow, reducing cardiac contractility (e.g., beta-adrenergic or calcium channel blockers), or increasing heat storage (e.g., amphetamines, salicylates in larger doses). These effects contribute to hyperthermia and reduce exercise performance. Obviously, drug abuse may predispose a person to hyperthermia (19) and heatstroke (2).

- Skin disorders such as miliaria rubra or sunburn may reduce sweat production by occluding epidermal pores of eccrine sweat glands (20, 21). These conditions are acquired and reversible (2, 22) and are discussed in chapter 7. Congenital dysfunction of sweat glands also is possible; examples include ectodermal dysplasia (2), chronic anhidrosis (21), psoriasis, and scleroderma (9).

- A febrile illness involves elevated body temperature that is independent of exercise. The metabolic heat produced during exercise increases the temperature of deep body tissues beyond that caused by a fever and may result in heatstroke (7). The presence of an intestinal infection, such as gastroenteritis, predisposes a person to heatstoke (1, 23). This heat intolerance is transient and may last up to 12 weeks (16).

- Poor cardiorespiratory physical fitness reduces the capacity to respond to the demands of exercise in a hot environment. In studies of U.S. soldiers (1), Belgian mine workers (24), and Israeli soldiers (25), subjects with a maximal aerobic power of less than about 40 ml $\cdot$ kg$^{-1}\cdot$ min^{-1} were adversely affected by exercise in the heat. Further, Fruth and Gisolfi (26) utilized a rat model of heatstroke pathophysiology to evaluate the effects of treadmill endurance running (up to 90 min/d) in a cool environment (23°C, 73.4°F) on responses to an exercise–heat stress challenge. The results of these laboratory tests are shown in figure 8.2. The endurance-trained animals (n = 29) ran longer in the heat, sustained greater thermal loads, and were less susceptible to exercise-induced thermal fatality than were sedentary control rats (n = 34) of a similar weight. Highly trained endurance athletes exhibited similar characteristics during exercise–heat tolerance tests, even though all of their training sessions were conducted in cool conditions (27).

- Inadequate heat acclimation reduces the ability to work or exercise in a hot environment. Three fundamental cardiovascular adaptations occur during 8 to 14 days of heat acclimation: (a) an increased maximal cardiac output, (b) a decreased peak heart rate in response to a given work load, and (c) an increased stroke volume. Other adaptations include increased plasma volume, increased sweat production, reduced body temperature, and reduced sodium concentration in sweat and urine (28).

- An investigation regarding former heatstroke patients surveyed the incidence of 20 potential predisposing factors (table 8.1). Sleep

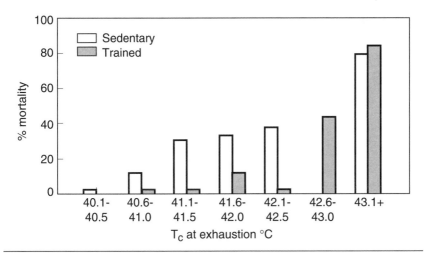

Figure 8.2 Percent mortality in rats, 24 hours after an exhausting exercise–heat exposure, plotted against rectal temperature at exhaustion. Lower mortality was observed among physically trained rats, unless their internal body temperature exceeded 42.5°C (108.5°F).

Data from Fruth and Gisolfi, 1983.

loss was the most common factor, experienced by 7 out of 10 individuals prior to their heatstroke episode (1). It also was implicated as a predisposing factor in heatstroke deaths that occurred in Israel (3, 23). Sleep loss has been shown to decrease skin blood flow and sweat rate at a given body temperature (29). It also recently has been shown to cause acute, relative insulin resistance (30). Because glucose uptake by skeletal muscle is essential to maintain prolonged exercise, this mechanism may result in fatigue associated with heatstroke.

• Uniforms worn by firefighters or soldiers (e.g., long-sleeved cotton uniform, rubberized protective suit with hood and gloves) and athletes (e.g., American football gear including helmet and protective pads) insulate the body and trap heat in all organs. In some cases (e.g., dieters, wrestlers), impermeable garments are worn to voluntarily induce weight loss as water.

• Adults who are obese exhibit compromised cardiac function (31) and have a lower tolerance of exercise in the heat (32, 33). The specific heat also may be markedly lower for these individuals than for subjects who are lean (34). Further, the energy cost of moving a large mass results in increased metabolic heat production. The elevation of body temperature is higher, for a given

Table 8.1

Host or Environmental Factors That Existed in the Days Before Exertional Heatstroke

Predisposing factor or warning signal	Number of exertional heatstroke patients (10 total) who experienced this factor
Sleep loss	7
Generalized fatigue	6
A warning sign of impending illness	6
A long exercise bout or workout	5
A long heat exposure (e.g., mowing grass, physical training)	5
A climatic heat wave	4
Reduced sweat secretion	3
Fever or disease	3
Dizziness, light-headedness	2
Dehydration	1
Taking medication (e.g., anti-histamine)	1
Excessive use of alcohol	1
Excessive use of caffeine	1
Consumption of a low-salt diet	1
Previous heat illness	1
Sunburn or skin rash	1
Immunization or inoculation	0
Use of diuretics	0
Previous difficulty with exercise in the heat	0
Diarrhea or vomiting	0

Reprinted, by permission, from L.E. Armstrong, J.P. DeLuca, and R.W. Hubbard, 1990, "Time course of recovery and heat acclimation ability of prior exertional heatstroke patients," *Medicine and Science in Sports and Exercise* 22(1): 36-48.

heat load and per kilogram of body mass, in individuals who are obese than in those who are not. As such, individuals who are obese have an increased risk of becoming hyperthermic during exercise.

• Large, muscular individuals also may be at increased risk of hyperthermia because metabolic heat is produced by muscles during exercise. This effect is exacerbated when heat dissipation is compromised by humid air or clothing that insulates the skin.

Predisposing factors for heatstroke differ among various populations. That is, the factors that increase susceptibility to heatstroke in young soldiers or athletes may differ from those that increased susceptibility in the religious pilgrims traveling to Mecca (1). The former group often consists of highly trained, motivated individuals exercising at high intensities in the heat.

Recurrent Exertional Heatstroke Due to Illness

Case Report 8.1

An acquired host factor may cause temporary heat intolerance and may make routine exercise in a hot environment difficult or a threat to health. This explains why some cases of exertional heatstroke involve athletes, soldiers, or laborers who previously performed a given activity many times in similar environments before experiencing life-threatening hyperthermia. This report describes a case of recurrent heatstroke that was attributed to a viral illness.

J.D. was a 19-year-old Israeli farmer who acclimated to heat during outdoor labor. He undertook a military training course that required daily strenuous activities. During the final 4 days of the course, J.D. experienced gastroenteritis, mild fever, diarrhea, and dehydration. After a vigorous 8-kilometer (5-mile) march, while wearing a 30-kilogram (66-lb) backpack, he collapsed and lost consciousness. The air temperature was 26°C (78.8°F) with 60% relative humidity. Because J.D. collapsed and exhibited fatigue and diarrhea, medical personnel administered a saline infusion and transported him to a local clinic. Fifteen hours after this incident, J.D. was awake with signs of dehydration, no neurological sequelae, a pulse of 100 beats/min, blood pressure of 100/65 mmHg, a normal chest X ray, and a normal electrocardiogram. All blood tests were normal except the serum enzymes AST (293 IU), LDH (506 IU), and CPK (668 IU). Normal values are 30, 350, and 100 IU, respectively. A diagnosis of exertional heatstroke was made on the basis of serum enzyme elevations and loss of consciousness. J.D. rested at home for a few days and then returned to regular duties with his military unit.

Three weeks following the aforementioned event, J.D. again participated in a vigorous march, in ambient conditions of 32°C

(89.6°F) and 20% relative humidity. Following a 20-kilometer (12.4-mile) march, he became confused and irritable, vomited, collapsed, and experienced generalized convulsive seizures. Upon admission to the hospital, one hour after collapsing, J.D. was alert and had a rectal temperature of 39°C (102.2°F), a heart rate of 120 beats/min, and blood pressure of 120/80 mmHg. Because 60 minutes had elapsed before hospital admission, J.D.'s rectal temperature likely was more than 40°C (104°F) at the time of his collapse. In the emergency department, J.D. vomited and experienced a second seizure. Rectal temperature returned to normal within the first day. Serum enzymes were elevated (AST, 350; LDH, 1447; CPK, 1252), but blood clotting factors and microorganism cultures were normal. The diagnosis was exertional heatstroke. After a 4-day recuperation in the hospital, J.D. was released when all laboratory values were found to be normal.

One month and five months after the second exertional heatstroke episode, J.D. visited an environmental physiology laboratory for two evaluations of his exercise–heat tolerance via bench stepping (12 steps/min; 3 h; 40°C, 104°F; 50% relative humidity). During the first exercise–heat tolerance test (August), he completed the 3 hours of exercise with a rectal temperature of 39.2°C (102.6°F) and maximal heart rate of 160 beats/min. J.D. finished the second evaluation (December) with a rectal temperature of 38.5°C (101.3°F) and maximal heart rate of 142 beats/min. During both tests, blood coagulation factors and enzyme levels were normal and whole-body sweat rates were similar. Observations of this second exercise–heat tolerance evaluation indicated greater heat tolerance than the first. This led investigators to conclude that J.D. was predisposed to the first heatstroke by an infectious disease (i.e., gastroenteritis with fever). Further, they surmised that either compromised thermoregulatory function or a continuous effect of the viral illness predisposed J.D. to the second heatstroke. A long period of rest was required to bring about complete recovery and normal exercise–heat tolerance. (This report was revised and condensed from reference 16.)

EXERTIONAL HEAT EXHAUSTION

Heat exhaustion is described in detail in chapter 4 and is defined here as the inability to continue exercise in a hot environment. Two types are classically described: water depletion heat exhaustion and salt depletion heat exhaustion. Because both types involve fluid volume depletion and may involve an electrolyte abnormality, any condition or factor that disrupts fluid–electrolyte homeostasis—including a large sweat or urine loss, diarrhea, vomiting, and consumption of a large volume of hypotonic or hypertonic fluid—may be considered a predisposing factor for heat exhaustion. By definition, exercise and ambient thermal stress also must be considered predisposing factors. Finally, the field observations of Hubbard, Gaffin, and Squire (9) indicate that approximately 20% of people with heat exhaustion have some form of viral or bacterial gastroenteritis.

Heat acclimation markedly affects fluid–electrolyte losses. In addition to decreased heart rate and rectal temperature, repeated days of heat exposure result in expanded plasma volume, increased sweat rate, and decreased sodium and chloride concentrations in sweat and urine (28). Clearly, unacclimated individuals will exhibit a different fluid–electrolyte profile following 8 to 14 days of exercise–heat exposure.

Two case-controlled investigations have evaluated the predisposing factors for heat exhaustion at 10-kilometer (6.2-mile) and 15-kilometer (9.3-mile) summer road races that were conducted in Sydney, Australia (35) and Atlanta, Georgia (36), respectively. The former study statistically identified motivation to exceed previous performance targets, failure to drink fluids during the run, failure of trained runners to acclimate to the heat, and a previous history of heat exhaustion as predisposing factors. The latter study identified two predisposing factors: poor prerace preparation and making an overly ambitious attempt to run faster.

HEAT SYNCOPE

As noted in chapter 4, syncope is a difficult disorder to predict and prevent because many medically serious and nonserious mechanisms can result in a temporary loss of consciousness (4). Heat syncope, however, is precipitated by long periods of standing or physical activity in a hot environment (37).

The classic predisposing factor for the greatest incidence of heat syncope involves unacclimated soldiers or marching band members standing at attention while wearing a uniform on the first hot day of the season. These individuals do not have the cardiovascular benefits of regular exposure to heat. Lack of heat acclimation is a predisposing factor for heat syncope (38). Thus, heat syncope is more likely to occur (39) when individuals perform unaccustomed exercise or labor and when ambient temperature or humidity suddenly rises.

Standing quietly in the heat after exercise also results in an increased venous blood volume in the legs, arms, and skin (18). If cardiac filling pressure and stroke volume decline rapidly, a fall in cardiac output and blood pressure will occur, resulting in heat

Band members are at great risk for heat syncope. It is important for them to acclimatize themselves to the hot weather before marching season begins.

syncope. Recent human studies have demonstrated that immediately after a single bout of exercise, profound changes occur in the mechanisms that regulate arterial pressure. Postexercise hypotension is characterized by a persistent drop in systemic vascular resistance that may last nearly 2 hours (40). Clearly, upright posture is a predisposing factor for heat syncope, especially in athletes who cease exercise and stand in place.

Decreased blood volume from sodium and water deficits also can precipitate orthostatic hypotension leading to syncope. Large sweat losses, vomiting, or diarrhea caused by gastroenteritis are predisposing factors for syncope (41), as is heavy menstrual bleeding, and these factors can be exacerbated by a high ambient temperature and venous pooling of blood in limbs and skin. Other conditions that encourage syncope include prolonged fasting (causing hypoglycemia), cardiac disorders (e.g., coronary artery disease arrhythmia), and medications that affect the circulatory system by either inhibiting the sympathetic nervous system (e.g., clonidine hydrochloride, methyldopa), inhibiting arterial vasoconstriction (e.g., prazosin hydrochloride, terazosin hydrochloride), or inducing dilation of blood vessels (e.g., hydralazine hydrochloride, minoxidil) (41).

EXERTIONAL HEAT CRAMPS

The exact mechanism involved in heat cramps is not completely understood. However, heat cramps have been induced in a laboratory (42), and treatment of heat cramps is rapid and effective when an intravenous saline solution is administered (42-44) or when a 0.1% NaCl solution is taken orally (39, 45). Virtually no authority doubts the connection between heat cramps and sodium (i.e., salt) depletion. Therefore, those factors that induce a sodium deficit, including large fluid–electrolyte losses caused by diarrhea, vomiting, diuresis, and sweat loss, are predisposing factors for heat cramps.

Certain competitive and occupational situations also are predisposing factors for heat cramps. Among athletes, tennis players and American football linemen are especially prone to heat cramps. Steel workers, coal miners, sugar cane cutters, and boiler operators are the most common occupational patients (45). Thus, muscular fatigue may play a role in the etiology of heat cramps. Interestingly, the incidence of heat cramps among laborers may be seasonal. For example, more than 90% of the hospitalized patients in the Youngstown, Ohio, steel district were admitted between April and October of the years 1929 to 1934 (45). This suggests that strenuous labor in a hot environment interacts with seasonal climatic stress to create a dual- or multifactorial predisposing factor.

Because heat acclimation reduces the likelihood of heat cramps (39), inadequate exposure to heat must be considered a predisposing factor for heat cramps. This relationship has been reported in Israeli soldiers (5), Indian soldiers (46), laboratory test participants (42), construction crews (47), and steel mill workers (45).

It appears that consumption of a large quantity of pure water also is a predisposing factor for exertional heat cramps. This was first recognized by Moss (48) and Haldane (49), who reported that heat cramps were often seen when coal miners replaced copious sweat losses with large volumes of water. This conclusion was later corroborated by Leithead and Gunn (50), who published observations of sugar cane cutters in British Guiana. These laborers worked strenuously in hot surroundings, sustained large sweat losses (e.g., 4.8-9.0 L per 8- to 10-h work shift), and consumed pure water amounting to 4.5 to 14 imperial pints per work shift. Leithead and Gunn (50) named this phenomenon water intoxication, repeating the original use of this term by Moss (48). Talbott (45) attributed the etiology of heat cramps to salt depletion, whereas Ladell (42) accorded almost equal significance to water intoxication and sodium loss. In fact, these two potential mechanisms are not wholly separable (50). In view of these facts, a low-salt diet may be considered a predisposing factor for heat cramps (45).

EXERTIONAL HYPONATREMIA

As noted in chapter 6, hyponatremia technically is not categorized as a heat illness in the International Classification of Diseases (51). However, because exercise in a hot environment stimulates considerably greater water turnover (i.e., sweat losses and fluid intake) than a comparable effort in cold conditions does, exertional hyponatremia appears in this book because most cases occur during endurance events or heavy labor conducted during summer months (52). In recent years, numerous other authors (10, 18, 53-59) also have considered exercise and heat to be predisposing factors for exertional hyponatremia.

Because hyponatremia results from a fluid–electrolyte imbalance, most cases of this illness involve a combination of two predisposing factors (52). First, sodium (Na^+) or sodium chloride ($NaCl$, that is, table salt) losses are not replaced adequately in dietary food or fluids. Second, excessive hypotonic fluid is consumed. This usually occurs because only pure water is consumed. Further, the volume of pure water consumed need not be large (see table 6.8 on page 129) to be considered excessive.

During the past decade, fluid intake has been emphasized so strongly that some participants in endurance events have consumed excessive amounts of fluid before, during, and after competition or prolonged exercise and subsequently have developed clinical signs and symptoms of hyponatremia (60). These athletes participated in distance running events longer than 42 kilometers (26 miles) and triathlons lasting 7 to 17 hours (52, 60-64). Similarly, hyponatremia has been reported among hikers (53, 56) and soldiers (55, 57, 65). These reports indicate that exercise must last at least 4 hours (52, 66) for hyponatremia to be a risk. Further, because all reports of exertional hyponatremia to date involve weight-bearing exercise, there may be something physiologically unique about upright posture that predisposes athletes to this fluid–electrolyte disturbance. Physiologically, this could occur because changes in blood pressure alter arginine vasopresssin (AVP; also known as antidiuretic hormone, ADH) or other hormone responses.

At least four previous studies indicated that sex may be a predisposing factor for exertional hyponatremia. Table 8.2 presents these findings. Backer et al. (53), Davis et al. (64), Ayus, Varon, and Arieff (67), and Davis et al. (61) reported that female hikers and marathon runners were, by far, more likely than males to develop serious symptomatic hyponatremia (19 women versus 4 men). Speedy et al. (63) performed a statistical analysis of the plasma sodium concentrations of 330 triathlon finishers (3.8-km swim, 180-km cycle, 42.2-km run). Women had significantly lower plasma sodium concentrations (133.7

versus 137.4 mmol · L^{-1}, p =.0001, n = 38) than did men. Similarly, Davis et al. (61) evaluated the medical records of severely hyponatremic (<125 mEq · L) marathon runners (42.2-km distance) and found that 93% were female. They suggested that hormonal or behavioral differences may have existed. This sex effect also may be caused by the fact that women normally have a smaller volume of total body water than do men and therefore require a smaller volume of water to dilute extracellular sodium (see table 6.8 on page 129); this also would apply to men with a small body mass. However, table 8.2 shows that exertional hyponatremia occurs in a higher percentage of males than females in some populations. Therefore, a conclusive answer to this matter awaits further research.

Table 8.2
Percentage of Females and Males in Cases of Exertional Hyponatremia

Event, number of cases reported, reference	Females (%)	Males (%)
Hikers in the Grand Canyon, Arizona, U.S.A. (n = 7) (53)	86	14
42.2-km marathon (n = 9) (64)	89	11
42.2-km marathon (n = 26) (61)	93	7
42.2-km marathon (n = 7) (67)	71	29
Military recruits, U.S. Armed Forces, 1989-1999 (n ~ 209) (52)	15	85
Summary of case reports involving various activities (n = 57) (68)	60	40
Grand mean (± SE)	69 ± 12	31 ± 12

SE = standard error of the mean.

Table 6.8 describes another electrolyte-related host factor: sweat sodium concentration. Montain, Sawka, and Wenger (52) recently noted that sweat sodium levels typically range from 10 to 60 mEq · L^{-1}. For this physiological feature to be the sole cause of exertional hyponatremia, exceptionally high sweat sodium concentrations would be required (e.g., 61-81 mEq · L^{-1}). Such levels are rare among acclimated endurance athletes, who normally exhibit sweat values of only 10 to 30 mEq Na$^+$ · L^{-1} (69). But are high sweat sodium levels possible? Montain, Sawka, and Wenger (52) believe that they can be and cite the research of Virjens and Rehrer (70) regarding one case of symptomatic hyponatremia that occurred after 2.5 hours of cycling. Calculations of sodium levels suggested that this athlete's sweat contained at least 100 mEq $Na^+ \cdot L^{-1}$. Similarly, Smith and colleagues (71) have reported hyponatremia in a case of salt deficiency heat exhaustion (see table 2.1 on page 20) in a young soldier whose sweat sodium concentration was very high (81-103 mEq $\cdot L^{-1}$) due to cystic fibrosis (see Case Report 4.1). Therefore, it appears that some cases of symptomatic exertional hyponatremia may be caused by abnormally high sweat sodium concentrations. However, most cases involve both sodium losses in sweat and excessive water intake (52, 72, 73).

In marathons and ultraendurance events, running speed, which impacts time on the race course, is a predisposing factor. Slower runners apparently exercise at a lower intensity with a lower fluid requirement, have greater opportunity and time to consume fluid, and have an increased risk of "overdrinking" and hyponatremia (60, 62, 74). As an example, the multivariate analysis of Speedy et al. (63) indicated that running time was a significant predictor of plasma sodium concentration in ultratriathletes (n = 330). Similarly, a research team led by Davis (61) found a significant inverse relationship between postrace serum sodium level and the time elapsed before presentation at the emergency department. Figure 8.3 illustrates this relationship. In addition, they compared the race times of hyponatremic marathon (42.2-km) runners to all other finishers. None of the individuals who were hyponatremic (n = 21) completed the race in less than 4 hours, whereas 16% of the 19,978 total participants accomplished this goal.

Finally, three additional physiological predisposing factors for exertional hyponatremia have been proposed, but none has been satisfactorily substantiated. First, it is possible that the hormone AVP responds abnormally to fluid overload. Normally, AVP levels remain low when body water is excessive. However, some authorities have postulated that inappropriately high levels of AVP occur in some cases of exertional hyponatremia. In fact, one case (66) reported such an abnormal response (+460% despite fluid overload) in a young test subject. Although this phenomenon may occur, data involving 14 distance runners with symptomatic hyponatremia demonstrated that none experienced AVP elevations (73). Second, it is possible that the kidneys fail to function normally in some cases of hyponatremia (75). This has been proposed because no one can explain why the body fails to excrete the excess water that is ingested during exercise (61). Possible mechanisms include decreased glomerular filtra-

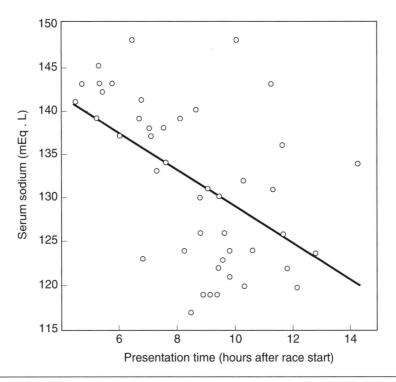

Figure 8.3 Inverse relationship between postrace serum sodium concentration (mEq · L^{-1}) and time elapsed (h) before presentation at a medical facility, in 46 marathon competitors. The linear regression prediction equation is y = 150.0 − 2.1(x), where y is serum sodium (mEq · L^{-1}) and x is time (hours), with r = 0.53 (p <.0002).

Reprinted from *Journal of Emergency Medicine,* vol. 21, Davis et al., "Exercise-associated hyponatremia," 47-57, © 2001, with permission from Elsevier Science.

tion rate or impaired electrolyte transport in nephrons (76). However, at least four studies have reported normal renal handling of fluid and electrolytes; these were cases of hyponatremic soldiers (57) and hyponatremic distance runners (73, 77). Third, nonsteroidal antiinflammatory drugs (NSAIDs) are known to potentiate the effect of AVP on the kidneys (i.e., at the collecting ducts) and have been associated with hyponatremia for many years (61, 78). This response involves impairment of water diuresis. A recent study (61) reported that three hyponatremic patients used NSAIDs during a marathon foot race. However, NSAID use was not measured in other studies. These three predisposing factors require further research before their validity can be established.

Virtually no authority cited here doubts that exertional hyponatremia involves retention of excess water. Inherent in this viewpoint are numerous reports of excessive drinking (53, 55, 59, 66, 75). This has prompted several experts to propose a reduction of the recommended fluid intake values for exercise in the heat (79) especially because the incidence of hyponatremia seems to have increased during the past 30 years (75, 79). However, among endurance athletes today, dehydration (i.e., weight loss) is much more common than weight gain, with a ratio of 22.6 to 1 (73). Therefore, reducing the consumption rate recommended by national sports medicine organizations (80, 81) does not seem reasonable, especially given the low incidence of hyponatremia (see table 2.2 on page 24).

The key to reducing the number of hyponatremia cases may lie in educating competitors about the ill effects of consuming too much water or hypotonic fluid. Often, competitors (61) and their friends (55) do not realize this medical reality. Davis et al. (61) eloquently stated the need to educate runners about developing rational fluid replacement strategies and about *optimizing* rather than *maximizing* fluid intake during extreme exercise. For example, the American College of Sports Medicine (80) and the National Athletic Trainers Association (81) recommend matching fluid intake with fluid lost in sweat and urine (see appendix A). Athletes and laborers can easily determine their sweat rate by measuring body weight before and after exercise, and correcting for fluid intake and urine production (81).

Considerable evidence indicates that hyponatremia occurs in individuals who "pound water" (i.e., drink excessively; see ref. 82), attempt to drink "as much fluid as possible" (61, 66), or believe that consuming water to excess prevents heat illnesses (55, 66). Coaches, educators, physicians, athletic trainers, dieticians, and race directors play a vital role in adjusting these behaviors.

SUMMARY

Many factors increase the risk of heat illness or exertional hyponatremia. These factors are distinct for each of the illnesses described and can be grouped into three categories. *Inherent host factors* include congenital sweat gland dysfunction, cystic fibrosis, hyperthyroidism, inappropriate hormonal responses, and large muscle mass. *Acquired factors* involve either functional impairment of thermoregulation (e.g., cardiovascular or infectious disease, inadequate nutrient intake, advanced age, obesity, sleep loss, moderate to severe dehydration, fluid excess, medications that alter heat dissipation, skin disorders in the miliaria family, lack of heat acclimation, lack of cardiorespiratory physical fitness) or residual injury (e.g., caused by disease, previous heatstroke, deep cutaneous burns). *Environmental factors* include ambient temperature, humidity, and solar radiation; insulated clothing; enclosed vehicles; and group, peer, or competitive pressure to perform exercise beyond one's capacity. Being aware of these factors and conducting preventive evaluations of each person who plans to exercise or work in a hot environment may spare individuals from serious illness or death.

Chapter 9

Considerations for the Medical Staff: Preventing, Identifying, and Treating Exertional Heat Illnesses

Douglas J. Casa, PhD, ATC, FACSM
William O. Roberts, MD, MS, FACSM

Exertional Heatstroke in a Short Road Race	Case Report 9.1

T.B., a 34-year-old male, completed an 11.4-kilometer (7.1-mile) seaside road race, dressed in running shorts and a singlet, and presented at the finish to the medical team supported by two medical triage personnel. The ambient temperature was 28°C (82°F) with a relative humidity of 79% under a hazy but sunny sky. On initial evaluation, T.B. appeared sweaty and ashen in color, and his skin was cool to the touch. He did not speak in response to questions and had a blank stare. His pulse was 110 and his rectal temperature topped out the clinical thermometer at 42.5°C (108°F). His stopwatch read 35 minutes. T.B. was immediately placed in an ice water tub for cooling. After 10 minutes in the tub, his rectal temperature measured 42°C (107.6°F). At 15 minutes, he became combative and had to be restrained in the tub. After 30 minutes of cooling, he began to interact with the medical team and could remember his name and midwestern city of residence. After 35 minutes in the cooling tub, his rectal temperature was 39°C (102°F), and he was removed from the tub for observation. His rectal temperature continued to fall to 36°C (97°F). After 60 minutes, his temperature was in the normal range, his mental status exam was normal, and his vital signs were normal. He was released to his hotel room with instructions to contact the medical team if he had further problems.

What separates this runner from other runners who collapse during or at the end of road races is his fast race pace in high-stress conditions, his high rectal temperature, and the loss of lower limb and central nervous system function. In addition, this runner benefited from a medical team that was prepared to evaluate and treat exertional heatstroke.

This chapter will prepare you to evaluate and treat athletes with exertional heatstroke and differentiate them from athletes who have less dangerous heat-related problems. Athletic participation or intense labor carry an ever-present risk of sustaining an exertional heat illness. The greatest risk occurs during intense activity in high heat stress environments. This chapter will focus on prevention, identification, and treatment of exertional heatstroke and the problems that occur more frequently in hot environments but are not caused by increased core body temperature: heat exhaustion, exertional heat cramps, heat syncope, and exertional hyponatremia. Many heat-related problems can be prevented by modifying the duration or intensity of practice or competition, and the morbidity and mortality of heat-related maladies can be minimized by prompt recognition and treatment.

RESPONSIBILITY OF THE MEDICAL STAFF

The role of the event medical staff is complex and challenging. Figure 9.1 shows the administrative areas that must be addressed by the medical team, although each factor may not apply in every athletic or work setting. Understanding the potential risks for developing heat-related illnesses, implementing practical policies and procedures that will reduce the occurrence of such disorders, and identifying and treating individuals affected by the heat will improve the safety profile of the athletic site. All of this takes pre-event planning.

REDUCING INCIDENCE
OF EXERTIONAL HEAT ILLNESSES

Prevention strategies can protect athletes from the adverse effects of heat. Explaining that the steps for heat illness prevention reduce heat illnesses and improve performance may generate acceptance of the prevention strategies among the participants, coaches, and administrators. Involving the athletes and administrators in the development of prevention strategies serves as a powerful educational tool for minimizing risk. Administrative actions such as scheduling the event in cooler times of the year, setting the starting times at cooler times of the day, and canceling or modifying events are out of the control of the participants yet are the most effective means of

Support services (ambulance, hospitals, police)

Recognition and treatment protocols

Previous event outcomes and epidemiology

Medical staff (MDs, ATCs, RNs, EMTs)

Venue or course aid stations

Parents or family (especially with youth events)

Equipment and supplies budget

Language issues

Event directors and administrators

Athletes or participants (number, fitness level, and so on)

Licensure and insurance

Record keeping

Event guidelines (wet bulb globe temperature: canceling, rescheduling, or altering)

Pre-event prevention

Adverse event protocol (media, and so on)

Officials (rules, policies, and so on)

Coaches (help implement policies, and so on)

Staff training

Spectator care

Figure 9.1 Factors to consider when coordinating medical coverage.

heat injury prevention. Successful use of personal strategies such as pre-event acclimation, fluid and salt replacement, adequate nutrition, satisfactory training, and sufficient sleep are under the direct control of the athlete and require strong educational efforts on the part of the administration, medical team, and coaches.

Identifying High Heat Stress Environments

A measure of heat stress is needed to determine the **thermal load** in an exercise environment. Heat dissipation must match endogenous and exogenous heat gain for an athlete to continue exercising with a normal core body temperature. When the thermal load exceeds the capacity to dissipate heat, hyperthermia ensues (1, 2). Factors that determine the thermal load include the intensity and duration of exercise, type and amount of clothing and equipment, environmental conditions (temperature, humidity, wind speed, direct and indirect radiation), and rest breaks. The medical staff should measure the environmental heat stress by assessing the wet bulb globe temperature (WBGT), the ambient temperature, heat and humidity, or dew point, and know how to interpret and utilize the findings from the associated charts or graphs.

A WBGT meter or calculation can quantify the heat stress. Table 9.1 provides an overview of WBGT readings and activity recommendations. If a WBGT meter is not available, the WBGT can be calculated using a hygrometer or sling psychrometer to measure the ambient and wet bulb temperatures and a black globe thermometer to measure the **black globe temperature** (see table 9.1). If a black globe thermometer is not available, an **ambient temperature** and relative humidity (or **wet bulb temperature)** chart can be used to predict the risk associated with exercise (3). If only ambient temperature and humidity can be ascertained, then a different risk scale, shown in figure 9.2, could be utilized. This figure shows the relationship between heat stress risk and humidity. Heat stress risk rises with increasing heat and humidity. This scale also factors in the role of various amounts of equipment and fluid/rest breaks. As temperature and humidity rise, more frequent fluid breaks should be scheduled. On bright, sunny days from mid-May to mid-September, you should add 5°F to the temperature between 10:00 A.M. and 4:00 P.M. (to account for radiant thermal load). Practices should be modified for the safety of the athletes to reflect the heat stress conditions.

Using only ambient temperature and relative humidity heat index discounts the radiant thermal load produced by the sun, which is significant during peak sun on a cloudless summer day. Artificial turf and blacktop surfaces exposed to direct sun can heat to 60°C (140°F), and this stored heat contributes to the radiant and conductive thermal load for the competitors. The **dew point** can also be used to estimate the heat stress, with dew points in the 60s considered stressful, in the 70s very stressful, and in the 80s dangerous for exercise. The medical, coaching, and administrative staff may be legally liable if standard heat modification recommendations are not utilized to minimize the risk of exertional heatstroke.

Preventing Significant Dehydration

Preventing dehydration is key to preventing exertional heat illnesses (4). Dehydration interferes with optimal thermoregulation and negates the positive effects of acclimation. In addition, fluid replacement during exercise decreases the likelihood and severity of most forms of heat-related illness (3-5). However, exertional heatstroke

Table 9.1
Categories of Risk for Exertional Heat Illnesses

Flag color	Level of risk	WBGT reading	Comments
Black	Very high risk	>28°C, 82°F	Incidence of exertional heatstroke and heat exhaustion is very high. Consider rescheduling, postponing, or canceling. Use extreme caution if activity will take place.
Red	High risk	23.1-28°C, 73.1-82°F	Use extra caution for athletes at high risk for heat illness. Warn medical staff and participants of high risk.
Yellow	Moderate risk	18.1-23°C, 65.1-73°F	Risk will increase if the temperature and sun exposure rises after the activity starts.
Green	Low risk	10.1-18°C, 50.1-65°F	Low risk exists for heat illness, but it can occur. Hypothermia is possible.
White	Lowest risk	<10°C, 50°F	This level carries the lowest risk of exertional heatstroke; heat exhaustion can occur; increased risk for hypothermia exists in slow or energy-depleted participants.

WBGT = wet bulb globe temperature.

WBGT (outdoors) = (Wet Bulb Temp × 0.7) + (Black Globe Temp × 0.2) + (Dry Bulb Temp × 0.1)

WBGT (indoors) = (Wet Bulb Temp × 0.7) + (Dry Bulb Temp × 0.3)

Adapted, by permission, from L.E. Armstrong et al., 1996, "American College of Sports Medicine position stand: Heat and cold illnesses during distance running," *Medicine and Science in Sports and Exercise* 28(12): i-x.

can occur without dehydration when endogenous heat production exceeds the cooling capacity (6-8). The signs and symptoms of dehydration include thirst, general discomfort, irritability, headache, weakness, dizziness, cramps, chills, nausea, orthostatic hypotension,

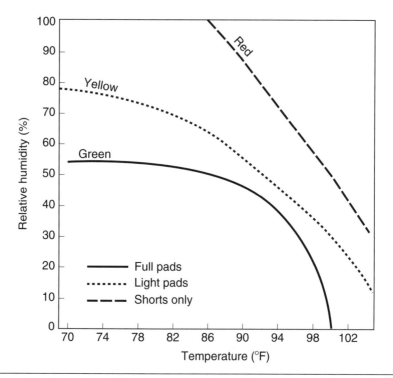

Figure 9.2 Heat stress risk temperature and humidity graph. Regular practices with full practice gear can be conducted for conditions that plot to the left of the green line. For conditions that plot between green and yellow, increase rest-to-work ratio with 5- to 10-minute rest and fluid breaks every 20 to 30 minutes. Practice should be in shorts; if for American football, can use helmets and shoulder pads (not full equipment). For conditions that plot between yellow and red, increase rest-to-work ratio with 5- to 10-minute rest and fluid breaks every 15 to 20 minutes. Practice should be in shorts with all protective equipment removed. When the temperature and relative humidity plot is to the right of the red line, cancel all practices, move practices into air-conditioned spaces, or hold them as walk-through sessions with no conditioning activities.

Kulka TJ & Kenney WL: Heat balance limits in football uniforms, *The Physician and Sportsmedicine* 2002; 30(7):29-39 © The McGraw-Hill Companies.

and orthostatic tachycardia (4, 5, 9). Performance progressively worsens with increasing dehydration, and endurance activities are negatively affected by as little as 1 to 2% dehydration (1, 4, 5, 10-13).

Education

Educational information in the form of flyers, brochures, or checklists should be given to participants. Information provided before the event should explain how to maximize performance in the heat while preventing heat-associated illness and dehydration. Information given on the day of the event should include specific information detailing the heat stress risks and prevention strategies. Postcare materials for heat-injured athletes should include follow-up plans, condition status changes that require reevaluation, and tips for recovery. The medical staff, event officials, and event volunteers also require instruction regarding heat illness prevention, detection, and management.

IDENTIFYING HYPERTHERMIA

An accurate estimate of core body temperature indicates the thermal stress to vital organs and is necessary for differentiating exertional heatstroke from other heat-related problems. The temperature of the skin (and shell) is influenced by the skin blood flow, sweat evaporation, ambient temperature, and relative humidity (14). Heat loss from the body is dependent upon the thermal gradient from the core to the shell to the environment. The greater the thermal gradient between the core, the skin surface, and the environment, the more efficiently heat is transferred by conduction to the environment (1, 2). If the ambient temperature is greater than the skin surface temperature, heat transfer will reverse and flow to the core. High relative humidity will decrease evaporation of sweat from the skin surface and impair the evaporative cooling mechanism (1).

The best method of estimating core temperature after exercise is a rectal temperature. Although the rectal temperature tends to lag behind the esophageal temperature, it provides a reliable core temperature in the field for medical classification and treatment (3, 6, 8, 15-21). Rectal temperature measures are inconvenient and embarrassing, but an accurate measurement of core temperature is necessary to accurately diagnose exertional heatstroke (14).

Oral, axillary, and tympanic (aural canal) measures, although convenient and modest, are inaccurate measures of core temperature after exercise in the heat (14, 18, 23, 24). Tympanic membrane (aural canal) thermometers measure a shell temperature rather than a core temperature in athletes who have been exercising and are often in the normal range when the actual core temperature is dangerously elevated (14, 22-24). In a recent study, the tympanic temperature measures were on average 2°C (4°F) lower than the rectal temperatures (24). After exercise, oral temperature measurement is largely influenced by respiratory rate, the temperature of rehydrating solutions, and the saliva temperature (14, 18). Axillary temperature is affected by sweat and by changes in the shell temperature (14).

DEFINING EXERTIONAL HEAT ILLNESS

The definition, clinical presentation, and diagnostic criteria for exertional heat illnesses are detailed in the following text and summarized in table 9.2. Exertional heat illnesses occur with much greater frequency in hot environments, but they occur in all temperature environments. Exertional hyponatremia can also occur in any environment, but it seems to be more frequent in the heat.

Exertional Heatstroke

The diagnosis of exertional heatstroke (EHS) is confirmed by a rectal temperature greater than 40°C (104°F) and the presence of central nervous system (CNS) dysfunction. EHS occurs when the thermoregulatory system cannot sufficiently remove the heat generated from muscle work, resulting in dangerous hyperthermia. EHS tends to occur during intense exercise in high WBGT environments. EHS also occurs when minimal rest or rehydration breaks are permitted during prolonged activity. EHS can lead to hypothalamic failure as the brain temperature rises to critical levels. Some of the common signs and symptoms of EHS include collapse during exercise, intense thirst, fatigue, impaired judgment, weakness, flushing, chills, hyperventilation, and dizziness. CNS depression signs are the most important marker of exertional heatstroke in a casualty with hyperthermia and include bizarre or irrational behavior, memory loss, loss of lower limb function and inability to walk, collapse, loss of consciousness, delirium, stupor, and coma. Seizures can occur but are less common in EHS that is recognized and treated early. The skin is usually ashen in color, because of circulatory collapse and impending shock, and it feels sweaty, wet, and often cool to the touch. Dry, hot skin is rare in EHS and implies hypothalamic failure (3, 6-8, 14-42). Altered CNS function and hyperthermia (rectal temperature >40°C, 104°F) are the two key characteristics that differentiate exertional heatstroke from heat exhaustion (3, 6, 16, 31, 45).

Heat Exhaustion

Heat exhaustion is characterized by the inability to continue exercise because of extreme fatigue and is probably a result of depletion of energy stores and dehydration. Heat exhaustion tends to occur during intense or prolonged exercise and is more prevalent in hot conditions. It is more common in participants who are not acclimated to heat and who do not rehydrate during the exercise. Some of the common signs and symptoms include headache, extreme weakness, syncope, profuse sweating, dizziness, nausea, "heat sensations" on the head or neck, feeling hot or cold, palpitations, and muscle cramping. The diagnosis is made in an athlete who collapses and cannot

Table 9.2
Identification and Treatment of Exertional Heat Illnesses

Exertional heat illness	Identification	Treatment
Exertional heatstroke (data from 3, 6, 7, 8, 14-43)	**Key history** • High-intensity exercise • High WBGT or high heat and humidity • Minimal rest breaks • Inadequate fluid replacement • Rapid onset • Second or third day of heat exposure • Sleeping in hot conditions • Athletic gear (such as football pads) and uniforms **Signs and symptoms** • Rectal temp >40°C (104°F) • CNS changes • Pulse 100-120 bpm • Systolic BP <100 mmHG • Pale and sweating skin • Dry skin if hypothalamic failure • Dizziness or weakness • General malaise **Diagnostic criteria** • Rectal temperature >40°C (104°F) • CNS dysfunction	**Acute** • Stop exercise. • Move to shade or air-conditioned shelter. • Elevate legs. • Cool immediately via ice water immersion or alternate protocol if immersion cannot take place (recognizing inferior cooling rates when not immersing). • Monitor core temperature, vital signs, and airway. • Stop immersion cooling at 38-39°C (100.4-102°F). • Insert an IV cather and replace fluid losses if dehydrated. • Control seizures. • Transport to hospital. **Post-acute** • Monitor for organ damage for 48-72 hours. • Rehydrate. • Monitor ECG changes. • Assess why heatstroke occurred. • Adhere to no activity in the heat for 7-14 days. • Have athlete sleep in air-conditioned space.

(continued)

Table 9.2

(continued)

Exertional heat illness	Identification	Treatment
Exertional hyponatremia (data from 4-7, 21, 25, 36, 44, 46-54)	**Key history** • Exercise >4 hours on successive days • Excessive fluid replacement • Excessive sweat losses replaced with water **Signs and symptoms** • Confusion • Extreme fatigue • Progressive headache • Nausea • Disorientation • Swelling of hands and fingers • Seizures • Dyspnea • Respiratory difficulty • Loss of consciousness • Dilute urine **Diagnostic criteria** • Sodium levels <130 mEq · L^{-1} • Rectal temperature <40°C (104°F)	**Acute** • Transfer to an emergency facility. • Insert IV for medication access. • Control seizures. • Determine mechanism of onset. • If water intoxication, treat with diuresis. • If sodium loss and dehydration, treat with fluid and saline replacement. **Post-acute** • Educate athlete about replacing fluid losses based on individual sweat rate. • Consume fluids that contain sodium. • Maintain normal meal pattern. • Include sodium in food and beverages. • Acclimate before events.

Heat exhaustion
(data from 3, 6, 8, 15, 16, 18, 21, 25, 30, 31, 37, 38, 40, 44, 45)

Key History

- Unacclimated
- Strenuous exercise
- Low fluid intake
- Low salt intake
- "Salty" sweat
- High WBGT and heat and humidity

Signs and symptoms

- Headache
- Extreme weakness
- Dizziness
- "Prickly heat" sensations
- Nausea or vomiting
- Profuse sweating
- Syncope
- Elevated HR
- Decreased BP

Diagnostic criteria

- Unable to continue exercising
- Rectal temperature <40°C (104°F)
- Minimal or no CNS dysfunction

Acute

- Have athlete rest supine with legs elevated.
- Rehydrate orally.
- Use IV rehydration if nausea or vomiting are present.

Post-acute

- Rest until symptoms resolved.
- Resume activity if medical evaluation is normal and activity is supervised.
- Watch closely for repeat episode.
- Resume activity the following day if well and unsupervised.
- Assess factors that caused the collapse.
- Institute prevention strategies.

(continued)

Table 9.2

(continued)

Exertional heat illness	Identification	Treatment
Heat cramps (data from 4, 6-8, 15, 16, 18, 21, 25, 30, 31, 44)	**Key history** • Excessive sweating • Intense exercise • Lower fitness level • Unacclimated • Rehydrating with hypotonic fluids • Near end of game or practice • High WBGT and heat and humidity **Signs and symptoms** • Intense pain and spasm in a muscle **Diagnostic criteria** • Intense muscle pain and spasm while exercising • No signs of heat exhaustion, exertional hyponatremia, or exertional heat-stroke • No acute muscle strain	**Acute** • Use neuroinhibition and prolonged stretch. • Rehydrate with salted fluids. • Use IV saline if athlete is nauseous or vomiting. • Use massage. **Post-acute** • Maintain hydration before and during activity. • Supplement salt in diet and rehydration fluids. • Eat nutritious meals. • Reassess training. • Follow acclimation procedures. • Have high-risk athletes consume extra sodium both before and during activity. • Utilize rehydration beverages that contain sodium.

Heat syncope
(data from 15, 16, 25, 30, 31, 44)

Key history
- Fainting
- Unacclimated exertion
- High WBGT and heat and humidity

Signs and symptoms
- Dizziness
- Tunnel vision
- Pale, sweaty skin
- Decreased pulse rate

Diagnostic criteria
- Brief fainting
- Normal rectal temperature
- Rapid recovery with lying down

Acute
- Elevate the feet above the level of the head.
- Rehydrate.
- Avoid sudden or prolonged standing.

Post-acute
- Avoid unaccustomed activity in the heat.
- Rise slowly after breaks in the heat.
- Avoid excessive stationary periods.
- Avoid sudden stationary periods after strenuous activity.

WBGT = wet bulb globe temperature; CNS = central nervous system; HR = heart rate; bpm = beats per minute; BP = blood pressure; mmHG = millimeters of mercury; ECG = electrocardiogram.

continue to exercise, is not hyperthermic (<40°C, 104°F), and has no true CNS dysfunction (3, 6, 8, 15, 16, 18, 21, 25, 30, 31, 37, 38, 40, 44, 45). Distinguishing exertional heatstroke from severe heat exhaustion is impossible without a rectal temperature measurement, and athletes with headache or severe muscle cramping should be evaluated for hyponatremia.

Heat Cramps

Exertional heat cramps are distinguished by severe involuntary muscle contractions caused by the triad of muscle fatigue, dehydration, and salt loss. Muscle cramping occurs in both hot and cold environments, but it is more prevalent in hot conditions.

Cramping is more likely to occur when excessive sweat losses are replaced by water alone (decreased sodium levels) as opposed to fluids containing sodium and may be more common in individuals who are not acclimated to heat or who have lower fitness levels than in those who are acclimated and highly fit. The examination will reveal no acute muscle strain yet intense painful spasm in a muscle that inhibits continued exercise (4, 6-8, 15, 16, 18, 21, 25, 30, 31, 44).

Heat Syncope

Fainting caused by vasodilation and pooling of blood in the lower body distinguishes this condition from other causes of collapse in the heat. Heat syncope presents when unacclimated individuals exert in the heat and faint from the vascular demands of exercise coupled with vasodilation in the legs and vagal inhibition of the heart rate resulting in insufficient vascular return to maintain cerebral blood flow. The signs and symptoms include dizziness, pale skin, decreased heart rate, weakness, and tunnel vision, ending in fainting. The athlete will recover quickly once the head is level with the legs and blood flow returns to the brain. The diagnostic criteria are a brief fainting episode and a normal rectal temperature, with the fainting episode being resolved by lying down (15, 16, 25, 30, 31, 44).

Exertional Hyponatremia

The diagnostic criteria for exertional hyponatremia are a serum sodium level of less than 130 meq $Na^+ \cdot L^{-1}$ and collapse, generally without hyperthermia (4-7, 21, 25, 36, 44, 46-54). The lowered serum sodium concentrations result from either excessive fluid intake that dilutes the total body sodium or excessive sodium losses in sweat replaced by water alone. Hyponatremia usually occurs in lengthy exercise sessions (>4 hours) when slower athletes ingest more fluid than they lose or aggressive athletes with high sweat rates and large sweat sodium losses do not rehydrate with a saline solution. The common signs and symptoms in the early stages include lightheadedness, dizziness, severe and progressive headache, and nausea. Athletes in the middle stages of hyponatremia may have vomiting, dyspnea, muscle cramps, confusion, and "puffiness" of the extremities and skin, yet normal blood pressure, pulse, and respiratory rates. In the late stages of hyponatremia, the individual will be ashen or gray in appearance and may experience obtundation (lowered level of consciousness), prolonged seizure, pulmonary edema, and cerebral edema. Published series of hyponatremia casualties show a mean serum sodium concentation of 121 ± 3 mmol $\cdot$ L^{-1} in obtunded patients with a range of 111 to 127 mmol $\cdot$ L^{-1}. The O_2 saturation may be decreased to less than 90% with impending pulmonary edema; the hematocrit will be decreased in overhydration and increased in dehydration.

Identification Summary

Exertional heatstroke and exertional hyponatremia are potentially fatal if not recognized and managed correctly (3, 6, 16, 37). A delay in administering appropriate care increases the likelihood of an untoward outcome. Hyponatremia and exertional heatstroke share many signs and symptoms and occur under similar circumstances (7, 21). Individuals with exertional hyponatremia most likely will have a rectal temperature less than 40°C (104°F) and a serum sodium level less than 130 meq Na $\cdot$ L^{-1} (7, 16, 21), whereas individuals with exertional heatstroke will be hyperthermic (rectal temperature >40°C,104°F) and have CNS dysfunction. Education of the medical and supervisory staff who advise and oversee individuals exercising in the heat can prevent a tragic outcome. Regular education sessions should address the signs, symptoms, preventive measures, and treatment of exertional heatstroke, heat exhaustion, exertional hyponatremia, and the other exercise-related causes of collapse noted in table 9.2.

TREATMENT OF EXERTIONAL HEAT ILLNESSES

A common misconception in the management of heat illness is that a collapsed athlete is always dehydrated (6, 8). The diagnosis of dehydration should be based upon clinical evidence of excess fluid losses and not assumed to be the cause of collapse. Administering fluids to an athlete who has collapsed from hyponatremia will worsen the condition and increase the risk of death (7). Similarly, exertional heatstroke patients often are not significantly dehydrated after intense activity of 30 to 60 minutes (6, 8). Marathon participants are more likely to experience exercise exhaustion, due to depletion of intramuscular fuels such as glycogen, than exertional heatstroke. Events that last 4 to 10 hours are associated with an increased risk of hyponatremia because competitors have the opportunity to drink too much fluid (21, 36). A correct diagnosis, even though it takes a few minutes, will improve care.

Exertional Heatstroke

The ultimate priority of EHS management is immediate cooling, because the degree and duration of hyperthermia in critical organs will impact the medical outcome (15, 16, 33). The best field method for rapidly cooling an exertional heatstroke casualty is ice water immersion. Immersion has superior cooling rates, can be done in the field if the athlete is stable, and may decrease the need for IV fluid replacement, because the hydrostatic pressure of the water will augment the blood pressure and vasoconstriction of the shell will shunt blood to the core (3, 25, 26, 28, 33, 55). Ice water (0°C, 32°F) and cool water (13-19°C, 55-60°F) immersion provide almost identical cooling rates for hyperthermic runners (27). During ice water immersion, expect a cooling rate of approximately 0.15-0.20°C per minute. If ice or cold water immersion is not possible, ice bags or ice water towels can be placed over the areas of greatest heat loss—the neck, armpits, groin, and behind the knees (no harm would be done placing ice bags in additional spots; this mode of cooling is effective during transport). Alternatively, cold towels can be placed over the chest, head, arms, and legs while a fan is blowing on the patient (3, 6, 19). Other methods of body cooling include sponging with cold, wet towels; using a warm or cool mist spray with fans; using fans alone; or utilizing a combination of these methods (15, 33).

The medical staff should establish a treatment area that allows ease of care. For example, the setup should ensure space to work on both sides of the tub to monitor vital signs, administer IVs, and remove the patient when cooling therapy is completed. The treatment area may be in a tent to provide shade and lessen some of the effects of the environment; if possible, an air-conditioned facility should be used. The staff should consider running fans in the treatment area to circulate the air. Tubs should be prepared in advance with water and ice so that they are ready for use when casualties arrive. When the diagnosis of exertional heatstroke is made, a rectal **thermistor,** if available, should be inserted, for continuous core temperature monitoring, before placing the casualty in the tub. Immersion cooling is labor intensive and may not be feasible for a single caregiver to manage alone, especially if the athlete becomes combative during cooling. An indwelling rectal thermistor is preferable to a rectal thermometer during cooling to allow constant monitoring of the core temperature. The individual to be cooled should be positioned so that the head is placed on one of the tub edges and the legs are resting in the tub or over the other edge of the tub (if the casualty will not fit in the tub). The more surface area of the arms, trunk, and legs that can be submerged in the water, the more rapid the conductive heat exchange for core body cooling. Continuous supervision is necessary because the patient could easily slip into the tub and occlude the airway. The team should have a bucket available for emesis and be prepared to roll the athlete in the tub to use the bucket if vomiting occurs. The heart rate, blood pressure, and rectal temperature should be measured and recorded every 5 to 10 minutes. Regular monitoring of core temperature prevents overcooling. Start an IV line in an arm, if access is needed for medication or fluid replacement. Remove the patient from the tub when the rectal temperature is 38-39°C (100.4-102°F), and continue monitoring rectal temperature and vital signs. The athlete should be transported to a medical facility unless the medical team is confident that the casualty was recognized and treated before any organ damage occurred.

Exertional Hyponatremia

An athlete with the clinical presentation of hyponatremia should be transferred to a hospital for correction of the sodium deficit or the fluid overload. Fluid volume should not be replaced unless dehydration is demonstrated in the athlete. If overhydration is present, natural diuresis may solve the problem, but the athlete should be monitored. If the athlete develops cerebral or pulmonary edema, loop diuretics and hypertonic saline may be needed to correct the electrolyte imbalance. Matching fluid replacement to sweat losses will prevent a recurrence of water intoxication. Additional salt in the regular meals will also be beneficial (48-54).

Heat Exhaustion

The treatment for heat exhaustion is supine rest in a shaded environment with the legs and buttocks elevated to improve core circulation. Cooling can also be used if it will make the athlete more comfortable. The athlete should rehydrate until he or she is able to urinate. If nausea or vomiting persists after 10 to 30 minutes of rest and leg elevation, consider IV hydration. After the acute episode of heat exhaustion has passed, exercise may be resumed if the athlete has been symptom free for 30 to 60 minutes. An athlete returned to intense exercise in the heat on the day of heat exhaustion must be observed closely for relapse and the possibility of sustaining exertional heatstroke (3, 6, 9, 21, 25, 37, 38).

Heat Cramps

The immediate treatment of heat cramps is prolonged stretch of the muscle and deep pressure over the area of spasm for neuroinhibition.

Sodium replacement with a saline solution will usually reverse the muscle spasm. Some fluids that are good sources of sodium include Pedialyte (Ross) with 45 mEq Na$^+$ · L^{-1} and Gatorade (18 mEq Na$^+$ · L^{-1}). For athletes with special sodium needs, these solutions may optimize sodium intake: Gatorlytes (Gatorade) with 75 mEq Na$^+$ · L^{-1} (with one Gatorlyte packet added to 20 oz of Gatorade), Rehydralyte (Ross) with 75 mEq Na$^+$ · L^{-1}, chicken bouillon broth, tomato juice, or two10-grain salt tablets dissolved in a liter of water. One of the simplest methods is adding salt (such as in the form of Gatorlytes) to a sports drink, because the flavoring of the sports drink may disguise the unfavorable taste of high amounts of sodium. This method assures known quantities of sodium (measure amount added) and provides water for rehydration. For example, a football player could easily lose 5 to 10 grams of sodium during a single day of two-a-day practices in the preseason. This loss requires special attention to the sodium content of the rehydration beverage and diet. If the athlete is nauseated or vomiting, intravenous normal saline solution can be used to replace sodium and water. Massage can provide some comfort for the muscle after the cramp is resolved and fluid is replaced. The medical staff and athletes should implement strategies to decrease the occurrence of muscle cramps. The primary prevention strategy is to supplement salt at meals for all athletes and during activity in known crampers for the first few days of practice in the heat. Other prevention strategies include avoiding dehydration with sports drinks, maximizing fitness, and acclimating to heat (4, 5, 6, 21, 25, 36, 44, 56-58).

Heat Syncope

If an athlete has fainted because of heat syncope, the athlete should lie in a supine position with the legs elevated above the level of the head, preferably in the shade. The medical staff should slowly transition the athlete from the supine position to a seated position, and finally to a supported standing position, before allowing unassisted walking. The athlete can return to activity when feeling well (15, 16, 25, 30, 31, 44).

STAFFING, SITE, EQUIPMENT, AND SUPPLY CONSIDERATIONS

Before the event, the medical director or medical committee will have to address the site-specific considerations that will allow the medical team to successfully manage heat-related problems. These administrative areas are listed in figure 9.1. The medical director must integrate the available data for the venue, including historical meteorologic conditions, activity intensity and duration, number of participants, previous event casualty rates, and anticipated fitness level (including acclimation) of the participants, to determine the site needs for staffing, equipment, and supplies (3, 13, 36, 40, 50, 59-67).

The medical team should be staffed by personnel who have the skills needed to treat the anticipated casualties. Physicians well versed in the care of heat- and exercise-induced medical casualties can supervise teams of volunteers to treat large numbers of collapsed athletes. Registered nurses, physician's assistants, and paramedics who are trained to start intravenous lines and administer intravenous solutions and medications are key personnel for events that anticipate treating casualties on-site. Intensive care unit and emergency room personnel are ideally suited for the high-intensity, high-volume care situations that occur in an event medical area. Emergency medical technicians (EMTs), certified athletic trainers (ATCs), and other first-aid trained volunteers can assist with care in the medical area, start first aid at the site of collapse, and perform triage on the sidelines or at the finish line of a large event. Sources of medically trained volunteers include the local hospitals and clinics, ambulance services, the American Red Cross, the state athletic trainers association, the National Ski Patrol, the Civil Bicycle Patrol, and Explorer Scout Troops.

Medical Sites

Medical personnel and first-aid stations should be located where participants have historically collapsed from heat-related problems. Comprehensive medical facilities should be located where the most casualties occur, such as the finish area or a particularly stressful section of a road race. Comprehensive medical facilities should be located at high-risk sites that have poor access to transportation, as delays in treatment can affect the outcome of exertional heatstroke. In some settings, mobile medical units equipped to care for heat injury and traveling with the participants may be a better alternative to multiple stationary teams. In industrial and large sport tournament settings, one central medical site will suffice if it is easily and quickly accessible for all in need of care.

Athletic Practices

A medical emergency plan to address the collapsed athlete should be established for all athletic practices at the high school, collegiate, and professional levels. Ideally, a certified athletic trainer should be present during games and practices, especially during high-risk heat situations such as two-a-day football practices in July and August. Many high schools and small colleges do not have athletic trainers at every practice or event, so the emergency plan should include phone access to the school nurse or local ambulance service for on-site evaluation and transportation to an emergency facility. A first-aid cooling protocol should be developed to begin cooling on-site for suspected exertional heatstroke. The coaches working in conjunction with the athletic trainer are responsible for ensuring that the training schedule includes adequate rest and fluid replacement breaks in hot weather and that football equipment is removed as the thermal load increases.

At-Risk Work Settings

In military settings, a medic or physician should be available when missions involve strenuous physical labor in the heat, and soldiers should be able to contact the medical staff whenever necessary. In industrial settings, the medical supervisor should be available, and workers should be able to easily contact this individual.

Games and Small-Scale Events

Football, soccer, lacrosse, and field hockey games and cross-country and long-distance track races held in hot conditions should have medical care on-site or have a plan of access to care if an athlete collapses. Even small-scale events in high heat stress conditions have the potential for a large number of heat-related problems, and in high-risk conditions an on-site medical team should be present to provide evaluation and care.

Large-Scale Events, High-Risk Situations

The American College of Sports Medicine (ACSM) (3) and International Amateur Athletic Federation (IAAF) (40) offer excellent general recommendations for medical care at large-scale road races from 5 kilometers through the marathon distance. Summer youth sport tournaments, especially soccer, and summer sports camps pose particularly high-risk situations, as children are more susceptible to the heat, may be supervised by untrained coaches, and may ignore temperature and fluid-replacement warnings. Events should be staffed based on the expected number and time span of heat-related cases. The staff for an 11.3-kilometer (7-mile) road race in hot conditions may evaluate and treat 50 to 60 heat-related problems in one hour, requiring a much larger staff than would a week-long soccer tournament with 10 to 20 heat casualties. For high-risk heat conditions, a team of four medical personnel using immersion cooling can care for one to two EHS casualties every 30 to 45 minutes. Ideally, each cooling team would have one physician, one registered nurse (RNs), one ATC, and one additional assistant for recording vital signs, although one physician could supervise two to three teams. A similar team of four can manage four to eight non-EHS casualties per hour. Reliable data from previous events in similar conditions can give a good estimate of casualties to set up an adequate number of care teams. As an example, the Twin Cities Marathon has 300 medical-related volunteers on the course and at the finish line for 6,000 to 8,000 participants. The finish area utilizes 15 to 20 physicians, 30 RNs, 50 EMTs, 20 ATCs, and 10 to 15 nonmedical support volunteers who take care of all the medical and trauma-related problems, including the heat-related problems, in the finish area. The physicians and nurses should be able to start intravenous lines, and several of the physicians and nurses should be using Advanced Cardiac Life Support (ACLS) protocol on a regular basis.

Figure 9.3 lists the equipment and supplies needed to treat the heat-related medical conditions discussed in this chapter. The amount of each supply should be estimated based on the anticipated number and type of casualties. Most events in hot conditions will require the basic tools of assessment. The decision to provide on-site care for heat illness will be based on the availability of qualified staff, the site accessibility, and the proximity to emergency services. Providing on-site care will increase the need for space and equipment.

Assessment Tools

A high-temperature rectal thermometer that measures up to at least 44°C (111°F) is essential to determine core temperature. A flexible rectal thermistor attached to a handheld temperature reader is preferable to a rectal thermometer for patients who require core temperature monitoring during immersion cooling treatment. A stethoscope, blood pressure cuff, and watch are needed to assess other vital signs.

Universal Precautions

Gloves, biohazard bags, sharps containers, disinfectant solutions and wipes, sanitizing soaps, a source of water (preferably a portable washstand), and biohazard clean-up kits are needed to prevent the spread of blood-borne pathogens.

Cooling Equipment and Supplies

The basic cooling supplies are tubs that will fit a body (small oval child's wading pools, inflatable rafts, stock feeding tubs, or large oval wash basins work well), a water source (bottled or from a tap), ice, storage ice chests, a garden hose, plastic bags for ice packs, fans, towels, and blankets. A tent, vehicle, or building will provide

Items

Stretchers or wheelchairs

Blankets

Bath towels

High temperature (43.3°C, 110°F) rectal thermometers

Adhesive tape (1 and 0.5 in.)

Disposable latex gloves

Blood pressure cuffs

IV fluid ($D_{5\%}$NS and NS)

Dextrose 50% in water

Sharps and biohazard disposal containers

Tables for medical supplies

Tubs for water immersion

Oxygen tanks with regulators, tubing, and masks

Rehydration fluids (water, CHO, sodium solution)

Glucose blood monitoring kits and strips

Injectable drugs (e.g., diazepam, magnesium sulfate)

Plastic or zipper-lock bags

Sodium analyzer

Cots

Gauze pads

Surgical soap

K-Y Jelly

Adhesive bandage strip

Stethoscopes

Intravenous (IV) administration kits

Tourniquets

IV catheters

Alcohol wipes

Water source

Fans for air circulation

Crushed ice in plastic bags

Cups

ALS ambulance

Garbage bags

Stopwatch or wrist watch

O_2 saturation analyzer

Other support considerations at medical stations

Generators

Toilet

Record keeping paperwork

Equipment and writing tables

Chairs

Portable washstand

Tent

Clipboards

Ambulance access

WBGT meter (sling psychrometer or hygrometer if not available)

Communication devices (walkie-talkies, cell phones, CBs)

Figure 9.3 Medical aid stations equipment and supplies for heat illness cases. This list was developed for large-scale running events, but may be modified to other events.

Adapted, by permission, from L.E. Armstrong et al., 1996, "American College of Sports Medicine position stand: Heat and cold illnesses during distance running," *Medicine and Science in Sports and Exercise* 28(12): i-x.

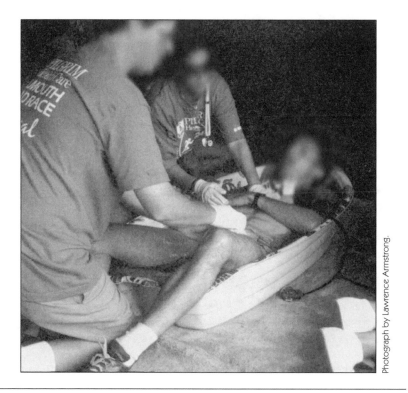

Photograph by Lawrence Armstrong.

Submerging an athlete in a tub full of water and ice is a good way to rapidly cool an athlete.

privacy and shade for the medical site. An air-conditioned building or vehicle helps with cooling.

Oral rehydration supplies include water, sports drinks, bouillon broth, fruit juices, ice, cups, straws, and coolers. If intravenous fluid replacement will be initiated on-site, IV solutions and administration sets will be required. Medications used to control muscle spasms and seizures (diazepam and magnesium sulfate) are administered intravenously and should be in the emergency medical kit. An electric power source or generator, work and equipment tables, cots, stretchers, wheelchairs, standard chairs, clipboards, pens or pencils, and medical record forms round out the equipment and supply list. Food and extra clothing for the patients are helpful. Food should be provided for the medical staff if the volunteer shift is longer than 3 to 4 hours. Volunteer food should be consumed away from the assessment and treatment areas.

Communication and Transportation

A medical plan that involves multiple stations or a large area must include a communication system for medical team members and emergency medical services. A dependable communication link with emergency services using a standard phone or cell phones is essential. Handheld radios and other communication devices can be used to keep the medical team connected. Back-up batteries and devices in case of equipment malfunction should be available. The medical plan also must include transportation for patients to the medical site and to the emergency facility. Ambulances, vans, cars, buses, golf carts, wheelchairs, gurneys, or litters can be used to move heat-injured athletes. If EHS is suspected, first aid with ice packing over the heat loss areas should not be delayed while waiting for transportation. Transportation to help staff move among venues also should be considered.

RECOMMENDATIONS FOR THE MEDICAL TEAM AND EVENT ADMINISTRATION

The following recommendations are essential for all medical team and event administration personnel.

- Understand the medical staff's responsibilities for the health of participants.
- Understand the influence of heat on exercise performance and physiological function.
- Implement prevention strategies to minimize risk of heat-associated illness.
- Develop a heat management plan for high-risk or dangerous heat conditions.
- Know the signs, symptoms, and clinical presentation of heat-related illnesses.
- Develop a medical team based on the number of competitors and risk associated with the event.
- Develop practical protocols to manage heat-related illnesses.
- Procure the equipment and supplies to assess and treat heat illnesses.
- Educate the medical staff, participants, and administrators regarding heat-related illnesses.

Figure 9.4 provides a checklist that can be used by medical staff working with athletic practices and competitions.

1. Pre-event preparation

_____ Challenge unsafe rules.

_____ Encourage athletes to drink according to individual fluid needs.

_____ Be familiar with athletes who have a history of a heat illness.

_____ Discourage alcohol, caffeine, and drug use before or during exercise.

_____ Encourage proper acclimatization procedures.

2. Checking hydration status

_____ Determine the preexercise weight of the athletes (when practical).

_____ Teach athletes to assess urine color and have urine color chart accessible.

_____ Help athletes determine their sweat rates to optimize the fluid replacement process.

_____ Have a refractometer accessible to check hydration status (to assess urine specific gravity at weigh-ins for at-risk individuals).

3. Environmental assessment

_____ Check the wet bulb globe temperature (WBGT) or temperature and relative humidity regularly during activity.

_____ Know the risk of a heat illness based on the WBGT or temperature and relative humidity.

_____ Develop plans for rescheduling events or practices during high-risk conditions.

4. Coach and athlete responsibilities

_____ Teach coaches and athletes the signs and symptoms of heat illnesses.

_____ Ensure proper hydration before the onset of activity.

_____ Make sure coaches allow rest and rehydration breaks.

_____ Adopt modifications to reduce risk in the heat.

_____ Ban rapid weight loss practices in weight class sports.

(continued)

Figure 9.4 When athletes exercise in the heat: A checklist for the medical staff.

Adapted, by permission, from D.J. Casa, 1999, "Exercise in the heat II: Critical concepts in rehydration, exertional heat illnesses, and maximizing athletic performance," _Journal of Athletic Training_ 34: 253-262.

---

5. **Event management**

_____ Ensure fluids will be available and accessible.

_____ Utilize carbohydrate and electrolyte drinks at events and practices lasting longer than 50 to 60 minutes and at events and practices that are extremely intense in nature.

_____ Rehydrate athletes to preexercise weight after an exercise session.

_____ Utilize shaded or indoor areas for practices when high temperatures and humidity.

6. **Treatment considerations**

_____ Become familiar with the most common early signs and symptoms of a heat illness.

_____ Have the proper field equipment and skills to promptly assess a heat illness.

_____ Have an emergency plan in place in case an immediate evacuation is needed.

_____ Have a tub available to initiate immediate cold or ice water immersion of exertional heatstroke patients.

_____ Have ice bags available for immediate cooling when ice water immersion is not possible.

_____ Identify shaded or air-conditioned areas for athletes to cool down, recover, and receive treatment.

7. **Other situation-specific considerations**

_____

_____

Figure 9.4 *(continued)*

Heat Exhaustion During Football Practice

Case Report 9.2

On the third day of two-a-day high school football practices, the temperature was 33.9°C (93°F) with a relative humidity of 70% during the mid-afternoon session. There was limited cloud cover and a 5- to 10-mile per hour breeze. B.P., a freshman offensive lineman, became unsteady and looked pale midway through the afternoon practice. B.P. reported feeling extremely fatigued and thirsty. The school athletic trainer suspected heat exhaustion and removed B.P. from practice. B.P. was rested, rehydrated, and observed in a shady spot within view of the football practice field. Cool towels were placed over his head, back, and legs for cooling and comfort.

The athletic trainer was proactive in reducing advanced exertional heat illness by observing the athletes closely and removing them from practice when they could no longer train safely and effectively. The athletic trainer utilized many other prevention strategies:

- Conducted weigh-ins before and after each practice to assess hydration losses and needs and did not allow participation if an athlete was dehydrated.
- Instituted regular rest and fluid breaks during practice.
- Mandated hydration during the breaks.
- Provided chilled sports drinks and water during breaks and practice.
- Reviewed the signs and symptoms of dehydration and exertional heat illnesses with the coaches and athletes.
- Allowed athletes to remove helmets and shoulder pads during intense conditioning drills.
- Identified high-risk athletes and implemented morning urine specific gravity assessments.
- Sent a hydration and heat illness education sheet home with each athlete to share with his parents.

Another prevention strategy to move the morning practice to an earlier time in the day and the afternoon practice to the evening, to minimize the thermal load, would decrease the incidence of heat exhaustion and risk of exertional heatstroke. Also, the coach and athletic trainer should develop a plan to transition into two-a-day practices and use of full equipment to maximize proper acclimatization and fitness levels over at least five days.

SUMMARY

Exertional heat illnesses can be reduced by instituting prevention strategies that address heat and humidity, length and intensity of practice, acclimation period, hydration during and away from practice, salt supplementation, appropriate work-to-rest ratios, and, in football, the gradual introduction of the full uniform and two-a-day practices during the early season conditioning and skill-building sessions. The medical staff must be vigilant during practice and competition in hot conditions, as exertional heatstroke can be missed without a high index of suspicion. Each site should have a heat emergency plan and the equipment needed to initiate treatment for heatstroke.

Chapter

Recommendations for Athletes and Weekend Warriors

Carl M. Maresh, PhD
Jaci L. VanHeest, PhD

Highly trained and experienced athletes must never underestimate the possibility of experiencing a heat illness during hot weather training sessions and competition, and they must take proper precautions to avoid such an illness. The combination of exercise and hyperthermia is one of the greatest stresses imposed on the human cardiovascular system and can present life-threatening challenges to thermoregulation in highly motivated athletes who drive themselves to extremes in hot environments (1). Therefore, the purpose of this final chapter is to present recommendations for adult athletes to reduce the risk of exertional heat illnesses and, thus, optimize exercise training and competitive efforts in the heat. Also included are recommendations for "weekend warriors," who are less adequately prepared for exercise in the heat than are competitive athletes.

Most athletes train on a consistent basis and regularly compete in sporting events. In contrast, weekend warriors are less physically fit, because of a less consistent training schedule, but periodically engage in intense competition. As such, these individuals can be highly motivated, with a level of fitness mismatched to their competitive aspirations. They may be unacclimatized or inadequately acclimatized to heat, have a greater risk of experiencing a heat illness, and pose a serious challenge to event organizers, athletic trainers, and health professionals.

To assess the risk of exertional heat illness, numerous inherited and lifestyle characteristics must be considered, including low cardiovascular fitness, obesity, lack of heat exposure, dehydration, sunburn, loss of sleep, fever, infection, alcohol excess, and a history of heat illness (4). These factors can be altered by changing lifestyle habits surrounding exercise and have been addressed in other publications (2, 3) and previous chapters of this book. The exertional heat illnesses that pose the greatest risk to athletes and weekend warriors include exertional heat cramps, heat exhaustion, and exertional heatstroke. Therefore, coaches, athletes, and event directors should know basic first-aid techniques for these illnesses (see chapters 3, 4, and 5).

It is just as important for these recreational athletes to take measures to prevent heat illnesses as it is for professional, competitive athletes.

The risk of heat illnesses can be reduced markedly by following preventive and precautionary measures. For example, the American College of Sports Medicine (ACSM) has provided guidelines to help runners prevent heat-related illnesses (4), and these guidelines can be applied to a wide range of athletic populations. Athletes should consider four periods of intervention: progressive heat acclimatization, the hours before exercise in the heat, the period of actual exercise, and the hours after exercise in the heat. The next four sections of this chapter discuss these windows of opportunity.

PROGRESSIVE HEAT ACCLIMATIZATION

Before competing in a hot environment, athletes should follow a heat-acclimatization program. This process is progressive, requires 10 to 14 days, and may be achieved by exercise in a natural environment (acclimatization), in an artificial environment (acclimation), or while wearing insulated clothing in a temperate environment (2, 3). Caution must be taken with this latter practice to avoid extreme hyperthermia (2). Heat acclimatization involves a unified complex of adaptations that include reduced heart rate, core body temperature, and sodium losses in sweat and urine; in addition, sweat rate increases. These adaptations dramatically reduce the incidence of heat exhaustion and heat syncope (see figure 4.4) because one effect of these adaptations is enhanced cardiovascular distribution of blood flow to vital organs.

By adhering to the following recommendations, athletes can optimize exercise–heat acclimatization before competition in a hot environment (2):

- Adequate fitness in cool environments must be attained before attempting to heat acclimatize. Enhanced cardiorespiratory physical fitness contributes to heat tolerance (5). Thus, the more fit a person is before beginning a heat-acclimatization program, the better able he or she will be to tolerate and benefit from the process.

- Exercise intensities during heat acclimatization should be greater than 50% $\dot{V}O_2$max. During the first 10 to 14 days of hot weather training, exercise should include gradual increases in intensity and duration (up to 90 min per day).

- High-intensity training should be performed during the cool, early morning or evening hours of the day. Less intense training should be undertaken during the hot midday hours.

- Fluid balance can be checked daily by measuring body weight upon rising each morning, using a reliable digital scale. A decrease of 2 to 3% in body weight should be offset with increased water consumption. Training duration and intensity should be reduced when there is a loss of 4 to 6% of body weight. If body weight is decreased by 7% or more, the athlete should consult with a physician who specializes in sports medicine and competitive athletes (3).

- Rectal temperature can be measured during or immediately after heat-acclimatization training to ensure that the final body temperature is within safe limits (i.e., no greater than 39.5°C,

103.2°F). The threshold for exertional heatstroke lies at core body temperatures of 39 to 40°C (102.2-104°F). To help prevent hyperthermia and heat exhaustion during training sessions, athletes should be aware of the environmental conditions and adjust training if needed, wear proper clothing, consume liberal amounts of fluids and electrolytes, and recognize the warning signs of heat illness (figure 10.1). Athletes also should exercise with a partner if possible.

Exertional heat illnesses	Warning signals
Heatstroke	Headache
	Lack of mental clarity
	Bizarre behavior
Heat exhaustion	Headache
	Agitation or irritability
	Unusual weakness or fatigue
	Dizziness
	"Heat sensations" on the head, neck, or arms
	Chills, goose flesh, or tingling
	Nausea or vomiting
	Unusual hyperventilation or breathlessness
Heat cramps	Very subtle, almost undetectable, muscle twitching
	Cramps in the abdominal, arm, or leg muscles

Figure 10.1 Warning signals of heatstroke, heat exhaustion, and heat cramps that athletes, coaches, and event directors can readily identify. For more warning signals, consult chapters 3, 4, and 5.
Data from 4, 7.

• Athletes who live in a cool environment, but who travel to compete in a hot environment, can induce partial acclimatization by wearing insulated clothing during training sessions in a mild environment. They should, however, leave some skin surface uncovered and monitor rectal temperature to avoid hyperthermia.

- The exercise–heat tolerance of children and adolescents may be less than that of adults, because of cardiovascular and thermoregulatory immaturity. A review by Armstrong and Maresh (6) identified pertinent questions for future research related to heat illness, acclimation, temperature regulation, exercise, and recovery in these younger populations.

THE HOURS BEFORE EXERCISE IN THE HEAT

After heat acclimatization has been achieved, day-to-day variations in heat tolerance may occur (7). Factors that reduce heat tolerance include dehydration, infectious disease, sleep loss, depletion of glycogen or glucose, medications, alcohol consumption, and a sudden increase in training intensity or duration. Not only do some of these reduce heat tolerance, but, as previously mentioned, some may contribute to a person's susceptibility to heat illness, regardless of heat-acclimatization status. If one or more of these factors is present, the athlete should scale back training on that day. Because many of these factors contribute to heat intolerance despite the presence of heat acclimatization, they should be monitored daily, as they are dynamic.

Both the ACSM (8) and the National Athletic Trainers' Association (NATA) (9) have published comprehensive guidelines regarding fluid replacement, including preexercise recommendations. Proper preexercise hydration is necessary for safe and effective training and competition in a hot environment. The following concepts focus on optimal hydration, with specific consideration to preexercise guidelines.

- Before exercise, the athlete should be well hydrated, with replenished muscle glycogen stores.
- Chronic dehydration may occur in athletes who perform repeated bouts of exercise on the same day or on consecutive days (10, 11). Athletes often require assistance in recognizing this problem. A prescribed, mandatory preexercise hydration regime may be useful. *Ad libitum* fluid consumption is usually insufficient to meet rehydration needs (12).
- Athletes should not attempt to exercise in the heat if they are ill, especially if they have a respiratory infection, diarrhea, vomiting, or fever (3, 4).
- Individuals should eat regular meals and a nutritionally balanced diet during the 24 hours before exercise, because a large portion of rehydration occurs during meals (11, 12). In most cases, water is an appropriate preexercise beverage because sufficient carbohydrate and electrolytes (e.g., sodium, chloride, potassium) are provided in a normal U.S. diet.

- Electrolytes help athletes regain and retain fluid–electrolyte balance after exercise-induced dehydration via sweating.
- Two to three hours before exercise, 500 to 600 milliliters of fluid should be consumed (8, 9). This provides ample time to urinate excess fluid.
- Ten to 20 minutes before exercise, 200 to 300 milliliters of fluids should be consumed (9).
- Body weight change, urine color, and **urine specific gravity** are valid indicators of hydration status (11, 13). Athletes, coaches, and trainers should routinely use these methods to assess preexercise hydration status.

DURING EXERCISE

In a hot environment, it is wise to exercise with a partner because the warning signals of heat illness (figure 10.1) can develop quickly. Training partners should pay attention to unusual behaviors or signs that might be warning signals of heat illness. Training partners should confide in each other if they feel unusual, weak, uncomfortably hot, or particularly stressed during a hot weather workout or competition.

Armstrong (7) has identified several guidelines for athletes, laborers, and soldiers to use in counteracting heat and humidity. Following these guidelines can minimize the physiological impact of hyperthermia and dehydration during exercise in the heat.

- Clothing should be lightweight, loose fitting, and porous to permit skin cooling via evaporation of sweat and dry heat loss.
- Cold water poured or sprayed on the skin or head provides a euphoric "high" in a hot environment, but it does very little to cool the body's inner organs (evaluated via rectal temperature) that are damaged by exertional heatstroke. The heat produced by exercising muscles is much greater than the amount of heat dissipated by a small volume of water evaporating from skin. Therefore, athletes should focus on ingesting cool water.
- Information regarding air temperature and humidity is available through television and radio weather reports. With this information, and by using available reference materials (7), an individual can determine if exercise intensity and duration should be reduced and the approximate risk of heat illness during exercise on a given day. Table 10.1 describes environmental conditions that are very stressful and perhaps dangerous for athletes who undertake intense or prolonged exercise.

Table 10.1

Environmental Conditions That Represent the Threshold of "High Risk" for Heat Illness Among Road Race Competitors[a]

Dry bulb temperature and relative humidity (rh)	WBGT[b]	Reference
	23°C or 73°F	15
23°C, 73°F with 100% rh	23°C or 73°F	4
27°C, 80°F with 66% rh		
31°C, 88°F with 30% rh		
34°C, 93°F with 20% rh		

[a]More stressful combinations of air temperature and humidity increase this risk to "very high."

[b]WBGT = wet bulb globe temperature; see references 4 and 15 for details.

- If one of the warning signals (figure 10.1) occurs, suggesting the onset of heat illness, the athlete should stop exercise, move to a shaded or air-conditioned area, cool the body if hyperthermia exists, consume fluids if dehydration exists, and seek medical attention.

- Heat illnesses occur when competition, peer pressure, or organizational discipline push healthy people beyond the point that they normally would stop exercise, rest, and consume fluids (16, 17). Athletes, coaches, and event directors should strive to limit such pressures and allow for adequate rest and hydration.

Proper hydration during exercise reduces cardiovascular strain, optimizes heat dissipation, and helps to maintain plasma volume and cardiac output (8, 14). Therefore, athletes are advised to heed these rehydration concepts.

- Fluids should be readily available, and athletes should consume as much as is comfortable, stopping regularly to drink. Understandably, this may not be easy to accomplish in sporting events that lack frequent breaks.

- Athletes should attempt to *optimize,* not maximize, fluid intake without overhydrating (see chapter 6). Athletes who know their rate of sweat production (i.e., body weight lost during exercise) can match fluid intake during exercise with sweat losses. This can be challenging to measure, but, with practice, athletes can learn to consume the correct amount of fluid (8).

- Exertional hyponatremia (i.e., serum sodium concentration <130 mEq · L^{-1}) occurs especially in endurance athletes (see chapter 6). This medical emergency results from overconsumption of water or other hypotonic fluid. Severe cases may involve coma, pulmonary or cerebral edema, and death (18, 19). The only certain way to avoid this life-threatening disorder is to consume an ample, but not excessive, volume of fluid.
- To optimize gastric emptying, 400 to 600 milliliters of fluid should be in the stomach at all times (8). If the fluid contains carbohydrates, the carbohydrate concentration should be between 4 and 8%. The rates of gastric emptying and intestinal absorption are similar for water and a dilute carbohydrate solution; however, the latter encourages fluid replacement and provides energy replenishment concurrently (20).
- In four situations, carbohydrate-electrolyte solutions are superior to water (21): (a) exercise is prolonged (>1 h); (b) exercise is strenuous (>70% $\dot{V}O_2$max); (c) either hyponatremia (see chapter 6) or salt deficiency heat exhaustion (see chapter 4) exists; and (d) within minutes of exercise cessation if immediate and optimal plasma volume recovery is necessary. For exercise sessions lasting less than one hour, water consumption should be sufficient. For sessions lasting more than one hour, a carbohydrate-electrolyte beverage (i.e., one containing sodium) is advisable.

THE HOURS AFTER EXERCISE IN THE HEAT

The period after exercise provides time to rest and replenish nutrients that have been depleted during exercise training or competition. If heeded, the following steps will help athletes prepare for subsequent training sessions or competitions.

- Proper rest after intense exercise or competition is particularly important when exercise is combined with an environmental heat stress. The body recuperates favorably if provided with adequate rest and sleep.
- Rehydration after exercise is critical and should be addressed independently of hydration before and during exercise (11). If the athlete was adequately hydrated at the start of exercise, reductions in body weight can be assumed to be body water lost during exercise.
- Body water can be replenished after exercise by drinking fluids with or without carbohydrates and electrolytes. Water alone, however, may not be the optimal rehydration beverage because it decreases osmolality, limits the drive to drink, and may in-

crease urine output. Salt (Na^+ or $NaCl$) in a rehydration beverage or diet will better conserve fluid volume and increase the drive to drink (21-23). For athletes who must rehydrate between exercise sessions without meal consumption, choosing a rehydration beverage with electrolytes is particularly important.

- The volume of fluid ingested during recovery should exceed the volume of sweat that was lost during exercise; otherwise, the athlete will remain somewhat dehydrated because of urinary losses. Shirreffs and colleagues (24) demonstrated that consuming a volume of fluid equal to 150% of weight loss, and utilizing fluid that contains sodium, optimizes bodily retention of fluid.

- Because replacement of expended glycogen stores in muscle and liver also is a postexercise goal, a rehydration solution should contain carbohydrates. Carbohydrate also may help to improve the intestinal absorption of sodium and water (25). However, the amount of carbohydrate in a sports drink (e.g., 19 g in a 591-ml, or 20-oz, serving) supplies only a small portion of the total carbohydrate content of a 2,000-kilocalorie diet (e.g., 325 g if the diet contains 65% carbohydrate). Therefore, athletes should incorporate adequate carbohydrate into their daily eating habits.

- Some athletes have used intravenous fluid infusion for rehydration between exercise sessions in the absence of clinical indications to do so. However, scientific studies (26, 27) that have assessed the efficacy of this practice reported no performance differences between intravenous and oral rehydration techniques when the rehydration and rest periods ranged from 20 to 75 minutes between exercise sessions. In terms of exercise performance, there are indications that oral rehydration may be superior, in subtle ways, to intravenous fluid replacement (26). It is possible, however, that future studies will find some combination of oral and intravenous rehydration optimal for exercise performance.

SUMMARY

Outdoor training and competitive plans should be altered in the presence of environmental warnings and host factors that predispose individuals to heat illness. Athletes should gradually increase their training duration and intensity in the heat, during a 10- to 14-day period, to allow the body to progressively adapt to exercise–heat stresses. Untrained individuals should not compete in the heat until they improve their fitness level. Fluid should be consumed before, during, and after training or competition in the heat. Body weight should be monitored daily, so that fluid intake can be optimized by matching the volume of sweat lost to the volume of fluid consumed. Overhydration is possible and, in cases of hyponatremia, may be life-threatening. Individuals should train with partners in hot environments so that the partners can watch for the warning signs of heat illnesses in each other. When temperature and humidity are severe, the duration and intensity of exercise should be reduced. Coaches, athletic trainers, and event organizers have a responsibility to guide athletes to train and compete safely in hot environments.

APPENDIX A

POSITION STANDS FROM THE ACSM AND THE NATA

American College of Sports Medicine Position Stand: Exercise and Fluid Replacement

Victor A. Convertino, PhD, FACSM (Chair)
Lawrence E. Armstrong, PhD, FACSM
Edward F. Coyle, PhD, FACSM
Gary W. Mack, PhD
Michael N. Sawka, PhD, FACSM
Leo C. Senay, Jr., PhD, FACSM
W. Michael Sherman, PhD, FACSM

This text appeared in *Medicine and Science in Sports and Exercise* Vol. 28, No. 1, pp. i-vii, 1996. Includes 92 references. To view the full text, visit www.acsm-msse.org/.

This abstract is reprinted, by permission, from V.A. Convertino et al., 1996, "Exercise and fluid replacement," *Medicine and Science in Sports and Exercise* 28(1): i-vii.

ABSTRACT

It is the position of the American College of Sports Medicine that adequate fluid replacement helps maintain hydration and, therefore, promotes the health, safety, and optimal physical performance of individuals participating in regular physical activity. This position statement is based on a comprehensive review and interpretation of scientific literature concerning the influence of fluid replacement on exercise performance and the risk of thermal injury associated with dehydration and hyperthermia. Based on available evidence, the American College of Sports Medicine makes the following general recommendations on the amount and composition of fluid that should be ingested in preparation for, during, and after exercise or athletic competition: 1) It is recommended that individuals consume a nutritionally balanced diet and drink adequate fluids during the 24-h period before an event, especially during the period that includes the meal prior to exercise, to promote proper hydration before exercise or competition. 2) It is recommended that individuals drink about 500 ml (about 17 ounces) of fluid about 2 h before exercise to promote adequate hydration and allow time for excretion of excess ingested water. 3) During exercise, athletes should start drinking early and at regular intervals in an attempt to consume fluids at a rate sufficient to replace all the water lost through sweating (i.e., body weight loss), or consume the maximal amount that can be tolerated. 4) It is recommended that ingested fluids be cooler than ambient temperature [between 15° and 22°C (59° and 72°F)] and flavored to enhance palatability and promote fluid replacement. Fluids should be readily available and served in containers that allow adequate volumes to be ingested with ease and with minimal interruption of exercise. 5) Addition of proper amounts of carbohydrates and/or electrolytes to a fluid replacement solution is recommended for exercise events of duration greater than 1 h since it does not significantly impair water delivery to the body and may enhance performance. During exercise lasting less than 1 h, there is little evidence of physiological or physical performance differences between consuming a carbohydrate-electrolyte drink and plain water. 6) During intense exercise lasting longer than 1 h, it is recommended that carbohydrates be ingested at a rate of 30-60 g · h^{-1} to maintain oxidation of carbohydrates and delay fatigue. This rate of carbohydrate intake can be achieved without compromising fluid delivery by drinking 600-1200 ml · h^{-1} of solutions containing 4%-8% carbohydrates (g · 100 ml · h^{-1}). The carbohydrates can be sugars (glucose or sucrose) or starch (e.g., maltodextrin). 7) Inclusion of sodium (0.5-0.7 g · L^{-1} of water) in the rehydration solution ingested during exercise lasting longer than 1 h is recommended since it may be advantageous in enhancing palatability, promoting fluid retention, and possibly preventing hyponatremia in certain individuals who drink excessive quantities of fluid. There is little physiological basis for the presence of sodium in an oral rehydration solution for enhancing intestinal water absorption as long as sodium is sufficiently available from the previous meal.

American College of Sports Medicine Position Stand: Heat and Cold Illnesses During Distance Running

Lawrence E. Armstrong, PhD, FACSM (Chair)
Yoram Epstein, PhD
John E. Greenleaf, PhD, FACSM
Emily M. Haymes, PhD, FACSM
Roger W. Hubbard, PhD
William O. Roberts, MD, FACSM
Paul D. Thompson, MD, FACSM

This text appeared in *Medicine and Science in Sports and Exercise* Vol. 28, No. 12, pp. i-x, 1996. Includes 97 references. To view the full text, visit www.acsm-msse.org/.

This abstract is reprinted, by permission, from L.E. Armstrong et al., 1996, "Position stand: Heat and cold illnesses during distance running," *Journal of Athletic Training* 28(12): i-x.

ABSTRACT

Many recreational and elite runners participate in distance races each year. When these events are conducted in hot or cold conditions, the risk of environmental illness increases. However, exertional hyperthermia, hypothermia, dehydration, and other related problems may be minimized with pre-event education and preparation. This position stand provides recommendations for the medical director and other race officials in the following areas: scheduling; organizing personnel, facilities, supplies, equipment, and communication; providing competitor education; measuring environmental stress; providing fluids; and avoiding potential legal liabilities. This document also describes the predisposing conditions, recognition, and treatment of the four most common environmental illnesses: heat exhaustion, heatstroke, hypothermia, and frostbite. The objectives of this position stand are: 1) To educate distance running event officials and participants about the most common forms of environmental illness including predisposing conditions, warning signs, susceptibility, and incidence reduction. 2) To advise race officials of their legal responsibilities and potential liability with regard to event safety and injury prevention. 3) To recommend that race officials consult local weather archives and plan events at times likely to be of low environmental stress to minimize detrimental effects on participants. 4) To encourage race officials to warn participants about environmental stress on race day and its implications for heat and cold illness. 5) To inform race officials of preventive actions that may reduce debilitation and environmental illness. 6) To describe the personnel, equipment, and supplies necessary to reduce and treat cases of collapse and environmental illness.

National Athletic Trainers' Association Position Statement: Fluid Replacement for Athletes

Douglas J. Casa, PhD, ATC, CSCS (Chair)
Lawrence E. Armstrong, PhD, FACSM
Susan K. Hillman, MS, MA, ATC, PT
Scott J. Montain, PhD, FACSM
Ralph V. Reiff, MEd, ATC
Brent S.E. Rich, MD, ATC
William O. Roberts, MD, MS, FACSM
Jennifer A. Stone, MS, ATC

This text appeared in *Journal of Athletic Training* Vol. 35, No. 2, pp. 212-224, 2000. Includes 220 references. To view the full text, visit www.nata.org/publications/otherpub/positionstatements.htm.

This abstract is reprinted, by permission, from D.J. Casa et al., 2000, "NATA position statement: Fluid replacement for athletes," *Journal of Athletic Training* 35(2): 212-224.

ABSTRACT

- **Objective:** To present recommendations to optimize the fluid-replacement practices of athletes.

- **Background:** Dehydration can compromise athletic performance and increase the risk of exertional heat injury. Athletes do not voluntarily drink sufficient water to prevent dehydration during physical activity. Drinking behavior can be modified by education, increasing accessibility, and optimizing palatability. However, excessive overdrinking should be avoided because it can also compromise physical performance and health. We provide practical recommendations regarding fluid replacement for athletes.

- **Recommendations:** Educate athletes regarding the risks of dehydration and overhydration on health and physical performance. Work with individual athletes to develop fluid-replacement practices that optimize hydration status before, during, and after competition.

National Athletic Trainers' Association Position Statement: Exertional Heat Illnesses

Helen M. Binkley, PhD, ATC, CSCS, NSCA-CPT (Chair)
Joseph Beckett, EdD, ATC
Douglas J. Casa, PhD, ATC, FACSM
Douglas M. Kleiner, PhD, ATC, FACSM
Paul E. Plummer, MA, ATC

This text appeared in *Journal of Athletic Training* Vol. 37, No. 3, pp. 329-343, 2002. Includes 230 references. To view the full text, visit www.nata.org/publications/otherpub/positionstatements.htm.

This abstract is reprinted, by permission, from H. Binkley et al., 2000, "NATA position statement: Exertional heat illnesses," *Journal of Athletic Training* 37(3): 329-343.

ABSTRACT

- **Objective:** To present recommendations for the prevention, recognition, treatment, and return to play for athletes with exertional heat illnesses and to describe the relevant physiology of thermoregulation.

- **Background:** Certified athletic trainers evaluate and treat heat-related injuries during athletic activity in "safe" and high-risk environments. While the recognition of heat illness has improved, the subtle signs and symptoms associated with heat illness are often overlooked, resulting in more serious problems for affected athletes. The recommendations presented here provide athletic trainers and allied health providers with an integrated scientific and practical approach to the prevention, recognition, and treatment of heat illness. These recommendations can be modified based on the environmental conditions of the site, the specific sport, and individual considerations to maximize safety and performance.

- **Recommendations:** Certified athletic trainers and other allied health providers should use these recommendations to establish on-site emergency plans for their venues and athletes. The primary goal of athlete safety is addressed through the prevention and recognition of heat-related illness and a well-developed plan to evaluate and treat affected athletes. Even with a heat illness prevention plan that includes medical screening, acclimatization, conditioning, environmental monitoring, and suitable practice adjustments, heat illness can and does occur. Athletic trainers and other allied health providers must be prepared to respond in an expedient manner to alleviate symptoms and minimize morbidity and mortality while ensuring rapid recovery.

Appendix B

The Body's Responses
to Various Heat Illnesses

The body is affected by the various heat illnesses in many ways. The table on pages 216-219 explains which systems of the body are affected by these heat illnesses:

- Heat stress
- Exertional heatstroke
- Exertional heat exhaustion
- Exercise-associated collapse
- Heat syncope
- Exertional heat cramps
- Exertional hyponatremia
- Minor heat illnesses

Systems Affected by Various Heat Illnesses

			Body systems		
Condition or illness	Central nervous system (CNS)	Cardiovascular system	Endocrine system	Excretory system	Other body systems
Exercise-heat stress (general, without signs and symptoms of illness; chapter 1)	Attempts to maintain homeostasis: • Hypothalamus receives afferent input • Hypothalamus sends efferent messages that maintain body temperature and regulate blood pressure	• Skin blood vessels dilate, dissipating heat to the environment • Heart rate, stroke volume, and blood vessel diameter change to regulate blood pressure	• Fluid-electrolyte hormones are released to retain water and NaCl	• Kidneys retain water and NaCl to offset sweat losses • Exercise in the heat decreases blood flow to kidneys	• Eccrine sweat glands secrete sweat, cooling the skin • Exercise in heat increases blood flow to active muscles and skin, but decreases blood flow to the organs

Exertional heatstroke (chapter 3)	• Hypothalamic temperature regulation is overwhelmed by large exercise-induced heat production • Mental acuity declines • Coma indicates advanced hyperthermia	• Increased blood flow to skin cools the body inadequately • Heart and blood vessels respond to maintain blood pressure • Levels of blood enzymes and coagulation factors may be increased	• Kidney function may decline due to rhabdomyolysis (muscle tissue damage) and decreased renal filtration	• Sweat evaporation cools the body inadequately • Vomiting or diarrhea may occur • Intestinal wall may become "leaky," allowing bacterial membrane fragments to enter the blood (endotoxemia) • All body organs, especially the liver, may be damaged by hyperthermia • May be fatal

217

Systems Affected by Various Heat Illnesses *(continued)*

Body systems

Condition or illness	Central nervous system (CNS)	Cardiovascular system	Endocrine system	Excretory system	Other body systems
Exertional heat exhaustion (chapter 4)	• CNS regulates the blood pressure decline caused by dehydration	• Heart and blood vessel responses compensate for fluid deficit and maintain blood pressure	• Fluid-electrolyte hormones decrease urine volume and sodium losses	• Kidneys conserve water and NaCl	• Exercise in the heat ceases • Vomiting or diarrhea may occur
Heat syncope (chapter 4)	• CNS attempts to regulate blood pressure to compensate for the effects of upright posture • Sensations of dizziness or nausea occur	• Blood pools in lower extremities and blood vessels dilate, causing hypotension • Syncope occurs	• Hormones are released to regulate blood pressure		

Condition	Nervous system	Cardiovascular/fluid	Fluid-electrolyte hormones	Whole-body water/sodium	Skeletal muscle/other
Exertional heat cramps (chapter 5)	• Dysfunctional CNS control of muscle tone and contractions may occur		• Fluid electrolyte hormones are released to compensate for sodium deficiency or water excess	• Imbalance of whole-body water and sodium occurs • Kidneys compensate for sodium deficiency	• Skeletal muscle cramps occur, likely due to altered sodium and potassium balance across the muscle cell membrane
Exertional hyponatremia (chapter 6)	• Mental acuity may decline • Coma occurs in severe cases • Cerebral edema is possible	• Baroreceptors in aorta and lungs sense increased pressure, due to fluid overload, and act to increase urine volume	• Fluid-electrolyte hormones are released to compensate for water excess	• Water excess stimulates the kidneys to regulate whole-body sodium balance	• Skeletal muscle cramps may occur • Pulmonary edema is possible • May be fatal
Minor heat illnesses (chapter 7)	• Cognitive performance may decline	• Interstitial edema may occur		• Imbalance of whole-body water and NaCl may occur	• Sweat glands may be obstructed • Inflammation may occur in skin

Exercise-associated collapse (EAC, chapter 4) involves a cluster of ailments that may occur in a cool or hot environment and may be due to numerous medical conditions. EAC does not include heart attack, chest pain, heatstroke, insulin reaction or shock, or other readily identifiable medical syndromes. This condition is characterized by abnormal body temperature (i.e., hypothermia or hyperthermia), altered mental status, muscle spasms, an inability to walk, or an inability to ingest oral fluids.

CNS = central nervous system, including the brain and spinal cord; NaCl = sodium chloride

GLOSSARY

ambient temperature—The temperature of environmental air.

anhidrosis—Diminished or absence of sweat gland secretion, possibly due to sunburn. May be local or widespread, temporary or permanent, congenital or acquired.

anhidrotic heat exhaustion—Extensive obstruction of sweat gland ducts leading to impaired evaporation, impaired cooling, and heat intolerance.

black globe temperature—Measured by inserting a dry bulb thermometer into a standard black metal globe; a component of the WBGT heat stress index.

capillary hydrostatic pressure—Pressure exerted by fluid within capillaries.

chyme—The mixture of partly digested food and secretions found in the stomach and small intestine during digestion of a meal.

conduction—Heat exchange between two objects that are in direct contact with each other.

convection—Heat exchange between an object and a fluid, such as air or water.

cyanosis—Slightly blue, gray, or dark purple discoloration of the skin, nail beds, lips, or mucous membranes caused by insufficient oxygenation of the blood.

cytokines—A protein that is involved in cell-to-cell communication and immune system responses (e.g., interleukin-6, tumor necrosis factor, interferon).

dermoepidermal junction—The interface between the outer skin layer (epidermis) and the deeper layer of skin (dermal layer).

dew point—The temperature at which vapor condenses into liquid or dew forms in nature.

dry bulb temperature—Standard air temperature that is measured with a thermometer; a component of the WBGT heat stress index.

dyspnea—Shortness of breath or "air hunger." Caused by deficiency of oxygen and excess of carbon dioxide in blood.

endotoxin—An intracellular biochemical substance that damages or kills body cells (e.g., bacteria inside the gastrointestinal tract). See also *lipopolysaccharide*.

euhydration—Normal hydration level.

euvolemic hyponatremia—An excess of total body water (fluid overload) with either normal or slightly reduced whole-body sodium.

evaporation—Heat exchange caused by water changing from the liquid phase to a gaseous phase.

exertional heat exhaustion—Inability to continue exercise in a hot environment.

fasciculations—Small localized muscle contractions, which can be observed through the skin and are caused by a number of fibers spontaneously discharging.

gram negative bacteria—A class of bacteria that do not retain a crystal violet stain in their outer membrane.

heat cramps—Painful spasms of skeletal muscles, usually after exercise in a hot environment.

heat edema—Swelling in body tissues during exposure to heat; fluid pools in the interstitial space; can be resolved with heat acclimation or relief from heat stress. Common among elderly individuals with poor circulation and in non-acclimatized individuals.

heat exhaustion—Inability to continue exercise in a hot environment.

heat syncope—A brief fainting episode in the absence of salt and water depletion, fluid loss, or hyperthermia, often subsequent to prolonged standing.

heatstroke—Whole-body hyperthermia that may lead to multi-organ damage; a medical emergency that requires rapid cooling therapy.

hemoconcentration—A relative increase in the number of red blood cells per unit of blood volume, due to the loss of plasma (i.e., the liquid portion of blood).

homeostasis—A dynamic equilibrium of the internal environment of the body. Maintained by feedback and regulation of the chemicals in tissues and body fluids.

hyperhydration—A short-term attempt to overhydrate; in healthy individuals, the kidneys remove this excess water.

hypervolemic hyponatremia—An excess of total body water (fluid overload), with an expanded extracellular fluid volume and increased whole-body sodium.

hypohydration—A steady-state body water deficit; in comparison, the term dehydration refers to an acute body water loss that leads to hypohydration.

hyponatremia—Sodium deficiency that involves a serum or plasma sodium concentration of < 130 mEq · L^{-1} (some authorities recognize < 135 mEq · L^{-1} as the definitive level). Is the result of the replacement of sodium losses with hypotonic fluid or pure water, loss of total body sodium, or both.

hypotonic—The osmotic concentration of a fluid is less than that of blood.

hypotonic hyponatremia—Low osmolality coexisting with low serum sodium.

hypovolemic hyponatremia—A reduced extracellular volume with deficits of total body sodium and water.

incidence—The frequency of occurrence of an illness over a period of time and in relation to the population in which it occurs.

interstitial hydrostatic pressure—Pressure exerted by fluid within the space between tissue cells.

interstitial oncotic pressure—Pressure exerted by interstitial proteins to drive fluid into the interstitial space.

interstitial space—Fluid-filled space between tissue cells.

intracorneal layer—Within the corneal skin layer.

intraepidermal layer—Within the epidermal skin layer.

keratin plugs—Keratin is a tough protein found in hair, nails, and the epidermis of the skin. In the skin, it forms flattened skin cells, which slough continually. These skin cells can form a plug of solid skin debris and clog sebaceous gland or sweat gland pore openings.

lipopolysaccharide (LPS)—A toxic portion of the bacterial cell membrane that stimulates an immune system response. It is released only when the membrane is destroyed; consists of lipid (fat) and carbohydrate. See also *endotoxin*.

mammillaria—Similar to miliaria rubra. Associated with anhidrotic heat exhaustion. Becomes more severe with increased exposure to heat and humidity. Involves pale, circular elevations on the skin, approximately 1 millimeter in diameter, between the neck and the waist. These "cobblestone" elevations appear with heat exposure and disappear with removal from heat.

mean corpuscular hemoglobin concentration—The hemoglobin concentration per unit volume of packed erythrocytes (red blood cells); calculated as (hemoglobin/hematocrit) × 100.

miliaria crystallina—A superficial form of miliaria that is rarely diagnosed. May be identified by the superficial eruption of clear, noninflammatory vesicles approximately 1 to 3 millimeters in size.

Newborns are particularly susceptible. Often associated with severe sunburns.

miliaria profunda—A deep form of miliaria. Firm, pale, or flesh-colored noninflammatory papules approximately 1 to 3 millimeters in size occur on the trunk and extremities. These papules are stimulated by sweat production and may become enlarged during physical activity or with exposure to extreme hot and humid conditions.

miliaria pustulosa—The small vesicles of miliaria rubra become pustular in nature and contain white or yellow purulent material. These superficial vesicles may contain bacteria.

miliaria rubra—Heat rash or prickly heat. A superficial, fine papular eruption and redness of the skin that follows excessive sweating and plugged sweat glands, particularly in hot–humid environments. Ordinarily confined to the torso, neck, and skin folds of the body. Often becomes infected or inflamed.

orthostatic hypotension—Low blood pressure that occurs because of an upright posture.

osmolality—Concentration of osmotically active particles in a solution.

osmotic forces—Forces exerted by ions that determine the direction of fluid movement between the different fluid compartments.

plasma oncotic pressure—Pressure exerted by plasma proteins to drive fluid into the plasma.

plasma volume—Liquid portion of the blood.

polyuria—Excessive secretion of urine.

radiation—Heat exchange in the form of electromagnetic waves.

salt deficiency heat exhaustion—Inadequate replacement of the sodium lost during sweating, leading to decreased extracellular and blood volumes.

set-point temperature—Regulated brain (hypothalamus) temperature. The set-point temperature changes with onset of fever.

splanchnic—Region of the torso including the gut, liver, and kidneys.

splanchnic vasoconstriction—Narrowing of blood vessels in the abdominal organs.

Starling forces—Factors that determine the direction of fluid movement across a capillary wall.

stratum corneum—The outermost layer of the epidermis.

subcorneal layer—The skin layer directly below the corneum.

syncope—Fainting.

thermal load—The amount of heat stored in body tissues and organs.

thermistor—An electronic device (e.g., a probe or a wire) that allows temperature to be measured digitally; based on the fact that electric current changes when metal is heated and cooled.

transient heat fatigue—Acute heat neurasthenia. Involves psychological symptoms such as extreme tiredness, disinclination to work, irritability, and errors in skilled performance, in the absence of salt and water deficiency and anhidrosis. Attributed to short-term exposure to extreme heat and poor ventilation.

tropical anhidrotic asthena—A chronic neurotic or depression-like illness, in the absence of a clearly defined organic heat disorder, attributable to prolonged exposure to a hot environment.

unspecified heat effects—Signs and symptoms not otherwise specified and caused by hot environments.

unspecified heat exhaustion—Occurs in the absence of salt and water deficiency and anhidrosis without other symptoms.

urine specific gravity—The mass of a urine sample compared with the mass of an equal volume of water.

water deficiency heat exhaustion—Attributable to inadequate water replacement in a hot environment. Characterized in the earliest stage by thirst, vague discomfort, anorexia, impatience, weariness, sleepiness, and dizziness. Restlessness, rapid breathing, cyanosis, and delirium eventually develop if left untreated.

water intoxication—Consumption of an excessive volume of water that results in symptoms such as headache, dizziness, vomiting, convulsions, coma, and death.

WBGT—wet bulb globe temperature; the most widely used heat stress index in industry and sports; may be used to assess the severity of hot environments; derived from a formula that incorporates the dry bulb, wet bulb, and black globe temperatures.

wet bulb temperature—A component of the WBGT heat stress index; measured with a water-saturated cloth wick over a thermometer (not immersed in water).

REFERENCES

Chapter 1

1. Rowell, L.B. 1986. *Human circulation regulation during physical stress*. New York: Oxford University Press.

2. Gagge, A.P., and R.R. Gonzalez. 1996. Mechanisms of heat exchange: Biophysics and physiology. In *Handbook of physiology*. Sect. 4, *Environmental physiology*, Vol. 1, edited by M.J. Fregly and C.M. Blatteis, 45-84. New York: Oxford University Press.

3. Santee, W.R., and R.R. Gonzalez. 1988. Characteristics of the thermal environment. In *Human performance physiology and environmental medicine at terrestrial extremes*, edited by K.B. Pandolf, M.N. Sawka, and R.R. Gonzalez, 1-44. Indianapolis: Benchmark Press.

4. Boulant, J.R. 1997. Thermoregulation. In *Fever: Basic mechanisms and management*, edited by P.A. Mackowiak, 35-58. Philadelphia: Lippincott-Raven.

5. Stitt, J.T. 1993. Central regulation of body temperature. In *Perspectives in exercise science and sports medicine*. Vol. 6, *Exercise, heat, and thermoregulation*, edited by C.V. Gisolfi, D.R. Lamb, and E.R. Nadel, 1-48. Dubuque, IA: Brown and Benchmark.

6. Sawka, M.N., C.B. Wenger, and K.B. Pandolf. 1996. Thermoregulatory responses to acute exercise-heat stress and heat acclimation. In *Handbook of physiology*. Sect. 4, *Environmental physiology*, Vol. 1, edited by M.J. Fregly and C.M. Blatteis, 157-185. New York: Oxford University Press.

7. Stephenson, L.A. 1997. Circadian timekeeping. In *Fever: Basic mechanisms and management*, edited by P.A. Mackowiak, 59-78. Philadelphia: Lippincott-Raven.

8. Vander, A.J., J.H. Sherman, and D.S. Luciano. 1990. *Human physiology*. New York: McGraw-Hill.

9. Simon, H.B., and G.H. Daniels. 1979. Hormonal hyperthermia. *American Journal of Medicine* 66: 257-263.

10. Valtin, H. 1983. *Renal function: Mechanisms preserving fluid and solute balance in health*. Boston: Little, Brown and Company.

11. Sawka, M.N., and E.F. Coyle. 1999. Influence of body water and blood volume on thermoregulation and exercise performance in the heat. In *Exercise and sport sciences reviews*. Vol. 27, edited by J.O. Holloszy, 167-218. Philadelphia: Lippincott Williams & Wilkins.

12. Dill, D.B., and D.L. Costill. 1974. Calculation of percentage changes in volumes of blood, plasma, and red cells in dehydration. *Journal of Applied Physiology* 37: 247-248.

13. Harrison, M.H. 1985. Effects of thermal stress and exercise on blood volume in humans. *Physiological Reviews* 65: 149-209.

14. Coyle, E.F., and M. Hamilton. 1990. Fluid replacement during exercise: Effects on physiological homeostasis and performance. In *Perspectives in exercise science and sports medicine*. Vol. 3, *Fluid homeostasis*, edited by C.V. Gisolfi and D.R. Lamb, 1-48. Carmel, IN: Benchmark Press.

15. Sawka, M.N., R.P. Francesconi, N.A. Pimental, and K.B. Pandolf. 1984. Hydration and vascular fluid shifts during exercise in the heat. *Journal of Applied Physiology* 56: 91-96.

16. Nielsen, M. 1938. Die regulation der korpertemperatur bei muskelarbeit. *Scandinavian Archives of Physiology* 79: 193-230.

17. Astrand, I. 1960. Aerobic work capacity in men and women. *Acta Physiologica Scandinavia* 49: 64-73.

18. Pawelcyzk, J.A. 1993. Neural control of skin and muscle blood flow during exercise and thermal stress. In *Perspectives in exercise science and sports medicine*. Vol. 6, *Exercise, heat, and thermoregulation*, edited by C.V. Gisolfi, D.R. Lamb, and E.R. Nadel, 119-178. Dubuque, IA: Brown and Benchmark.

19. Young, A.J. 1990. Energy substrate utilization during exercise in extreme environments. In *Exercise and sport sciences reviews*. Vol. 18, edited by K.B. Pandolf and J.O. Holloszy, 65-118. Baltimore: Williams & Wilkins.

20. Dimri, G.P., M.S. Malhotra, J.S. Gupta, T.S. Kumar, and B.S. Aora. 1980. Alterations in aerobic-anaerobic proportions of metabolism during work in heat. *European Journal of Applied Physiology* 45: 43-50.

21. Fink, W.J., D.L. Costill, and P.J. Van Handel. 1975. Leg muscle metabolism during exercise in the heat and cold. *European Journal of Applied Physiology* 34: 183-190.

22. Parkin, J.M., M.F. Carey, S. Zhao, and M.A. Febbraio. 1999. Effect of ambient temperature on human skeletal muscle metabolism during fatiguing submaximal exercise. *Journal of Applied Physiology* 86: 902-908.

23. Febbraio, M.A. 2000. Does muscle function and metabolism affect exercise performance in the heat? *Exercise and Sport Sciences Reviews* 28: 171-176.

24. Nielsen, B., J.R.S. Hales, S. Strange, K.J. Christensen, J. Warberg, and B. Saltin. 1993. Human circulatory and thermoregulatory adaptations with heat acclimation and exercise in a hot, dry environment. *Journal of Physiology* 460: 467-485.

25. Gonzalez-Alonso, J., C. Teller, S.L. Andersen, F.B. Jensen, T. Hyldig, and B. Nielsen. 1999. Influence of body temperature on the development of fatigue during prolonged exercise in the heat. *Journal of Applied Physiology* 86: 1032-1039.

26. Montain, S.J., M.N. Sawka, B.S. Cadarette, M.D. Quigley, and J.M. McKay. 1994. Physiological tolerance to uncompensable heat stress: Effects of exercise intensity, protective clothing, and climate. *Journal of Applied Physiology* 77: 216-222.

27. Sawka, M.N., A.J. Young, W.L. Latzka, P.D. Neufer, M.D. Quigley, and K.B. Pandolf. 1992. Human tolerance to heat strain during exercise: Influence of hydration. *Journal of Applied Physiology* 73: 368-375.

28. Wenger, C.B. 1988. Human heat acclimatization. In *Human performance physiology and environmental medicine at terrestrial extremes*, edited by K.B. Pandolf, M.N. Sawka, and R.R. Gonzalez, 153-198. Indianapolis: Benchmark Press.

29. Armstrong, L.E., and K.B. Pandolf. 1988. Physical training, cardiorespiratory physical fitness and exercise-heat tolerance. In *Human performance physiology and environmental medicine at terrestrial extremes*, edited by K.B. Pandolf, M.N. Sawka, and R.R. Gonzalez, 199-226. Indianapolis: Benchmark Press.

30. Armstrong, L.E., and C.M. Maresh. 1991. The induction and decay of heat acclimatization in trained athletes. *Sports Medicine (New Zealand)* 12: 302-312.

31. Convertino, V.A., L.E. Armstrong, E.F. Coyle, G.W. Mack, M.N. Sawka, L.C. Senay, and W.M. Sherman. 1996. American College of Sports Medicine position stand: Exercise and fluid replacement. *Medicine and Science in Sports and Exercise* 28 (1): i-vii.

32. Kenney, W.L. 1997. Thermoregulation at rest and during exercise in healthy older adults. In *Exercise and sport sciences reviews*. Vol. 25, edited by J.O. Holloszy, 41-76. Baltimore: Williams & Wilkins.

Chapter 2

1. Anonymous. 1967. A Roman experience with heat stroke in 24 B.C. *Bulletin of the New York Academy of Medicine* 43: 767-768.

2. Lindsay, P. 1936. *Kings of merry England*. London: Ivor Nicholson & Watson, Ltd.

3. Willcox, W.H. 1920. The nature, prevention, and treatment of heat hyperpyrexia. *British Medical Journal* 1: 392-397.

4. Malholtra, M.S., and Y. Venkataswamy. 1974. Heat casualties in the Indian Armed Forces. *Indian Journal of Medical Research* 62: 1293-1302.

5. Hubbard, R.W., W. Matthew, and D. Wright. 1982. *Survey and analysis of the heat casualty prevention experiment for Resphiblex 1-81, Operation Lancer Eagle, 43D, MAU (Report T 5/82)*. Natick, MA: U.S. Army Research Institute of Environmental Medicine.

6. Sutton, J.R., and O. Bar-Or. 1980. Thermal illness in fun running. *American Heart Journal* 100: 778-781.

7. Sutton, J.R. 1979. 43°C in fun runners! *Medical Journal of Australia* 2: 463-467.

8. Hart, L.E., B.P. Egier, A.G. Shimizu, P.J. Tandan, and J.R. Sutton. 1980. Exertional heatstroke: The runner's nemesis. *Canadian Medical Association Journal* 122: 1144-1149.

9. Bernheim, P.J., and J.N. Cox. 1960. Coup de chaleur et intoxication amphetaminique chez un sportif. *Schweiz Medicine Wochenschrift* 90: 322-331.

10. Nicholson, R.N., and K.W. Somerville. 1976. Heatstroke in a "run for fun." *British Medical Journal* 1: 1525-1526.

11. Aarseth, H.P., I. Eide, B. Skeie, and E. Thaulow. 1986. Heatstroke in endurance exercise. *Acta Medica Scandinavica* 220: 279-283.

12. Hanson, P.G., and S.W. Zimmerman. 1979. Exertional heatstroke in novice runners. *Journal of the American Medical Association* 242: 154-158.

13. Perlmutter, E.M. 1975. The Pittsburgh Marathon: 'Playing weather roulette.' *The Physician and Sportsmedicine* 14 (8): 132-134, 136-138.

14. Graber, C.D., R.B. Reinhold, and J.G. Breman. 1971. Fatal heat stroke. *Journal of the American Medical Association* 216: 1195-1196.

15. Sohal, R.S., S.C. Sun, and H.L. Colcolough. 1968. Heat stroke: An electron microscopic study of endothelial cell damage and disseminated intravascular coagulation. *Archives of Internal Medicine* 122: 43-48.

16. Leithead, C.S., J. Guthrie, and S. De La Place. 1958. Incidence, aetiology and prevention of heat illness on ships in the Persian Gulf. *Lancet* 1: 109-114.

17. Strydom, N.B., M.C. Kew, and M.E. Barry. 1966. Untitled. *Proceeding of the Mine Medical Officers Association* 46: 63-81.

18. Talbott, J.H. 1935. Heat cramps. In *Medicine,* edited by A.M. Chesney and D.L. Edsall, 323-376. Baltimore: Williams and Wilkins.

19. Weiner, J.S., and G.O. Horne. 1958. A classification of heat illness. A memorandum prepared for the climatic physiology committee of the medical research council. *British Medical Journal* 1: 1533-1535.

20. Minard, D. 1965. Nomenclature and classification of heat disorders. *Journal of the American Medical Association* 191: 150.

21. National Center for Health Statistics. 1998. *International Classification of Diseases, Ninth Revision, Clinical Modification (ICD-9-CM).* Hyattsville, MD: Author.

22. Callaham, M.L. 1979. *Emergency management of heat illness.* North Chicago, IL: Abbott Laboratories.

23. Epstein, Y. 2000. Exertional heatstroke: Lessons we tend to forget. *American Journal of Medicine and Sports* 2: 143-152.

24. Hubbard, R.W., and L.E. Armstrong. 1989. Hyperthermia: New thoughts on an old problem. *The Physician and Sportsmedicine* 17: 97-113.

25. Khogali, M. 1982. *Heat disorders with special reference to the Makkah Pilgrimage (The Hajj).* Saudi Arabia: Ministry of Health, Directorate of Health, Western Region.

26. Leithead, C.S., and A.R. Lind. 1964. *Heat stress and heat disorders.* Philadelphia: F.A. Davis.

27. Yarbrough, B.E., and R.W. Hubbard. 1989. Heat related illnesses. In *Management of wilderness and environmental emergencies,* edited by P.S. Auerbach and E.C. Geehr, 119-143. St. Louis: C.V. Mosby.

28. Gembitskiy, Y.V., G.N. Novozhilov, and S.D. Polozhentsev. 1985. Clinical aspects, diagnosis, and treatment of thermal injuries. *Voyenno - meditsinskiy Zhurnal* 3: 56-60.

29. Richards, D., R. Richards, P.J. Schofield, and J.R. Sutton. 1979. Management of heat exhaustion in Sydney's *The Sun* City-to-Surf fun runners. *The Medical Journal of Australia* 2: 453-457.

30. Richards, R., and D. Richards. 1984. Exertion-induced heat exhaustion and other medical aspects of the City-to-Surf fun runs, 1978-1984. *Medical Journal of Australia* 141: 799-805.

31. Hubbard, R.W., and L.E. Armstrong. 1988. The heat illnesses: Biochemical, ultrastructural, and fluid-electrolyte considerations. In *Human performance physiology and environmental medicine at terrestrial extremes,* edited by K.B. Pandolf, M.N. Sawka, and R.R. Gonzalez, 305-359. Indianapolis: Benchmark Press.

32. Shibolet, S., M.C. Lancaster, and Y. Danon. 1976. Heat stroke: A review. *Aviation, Space, and Environmental Medicine* 47 (3): 280-301.

33. Arday, D. 1990. Heat injury occurrence and prevention. Paper presented at symposium, Prevention and Treatment of Heat Injuries, April 24-26, 1990, Natick, MA, U.S. Army Research Institute of Environmental Medicine.

34. Convertino, V.A., L.E. Armstrong, E.F. Coyle, G.W. Mack, M.N. Sawka, L.C. Senay, and W.M. Sherman. 1996. American College of Sports Medicine

position stand: Exercise and fluid replacement. *Medicine and Science in Sports and Exercise* 28 (1): i-vii.

35. Hubbard, R.W., W.T. Matthew, and G. Thomas. 1983. *Bright Star 83 After Action Report (Specified Topics).* Natick, MA: U.S. Army Research Institute of Environmental Medicine.

36. Costrini, A.C., H.A. Pitt, A.B. Gustafson, and D.E. Uddin. 1979. Cardiovascular and metabolic manifestations of heat stroke and severe heat exhaustion. *American Journal of Medicine* 66: 296-302.

37. England, A.C., D.W. Fraser, A.W. Hightower, R. Tirinnanzi, D.J. Greenberg, K.E. Powell, C.M. Slovis, and R.A. Varsha. 1982. Preventing severe heat injury in runners: Suggestions from the 1979 Peachtree Road Race experience. *Annals of Internal Medicine* 97: 196-201.

38. Gumma, K., S.F. El-Mahrouky, N. Mahmoud, M.K.Y. Mustafa, and M. Khogali. 1983. The metabolic status of heat stroke patients: The Makkah experience. In *Heat stroke and temperature regulation,* edited by M. Khogali and J.R.S. Hales, 157-170. Sydney, Australia: Academic Press.

39. Costrini, A.C. 1990. Emergency treatment of exertional heatstroke and comparison of whole body cooling techniques. *Medicine and Science in Sports and Exercise* 22: 15-18.

40. Kark, J.A., P.Q. Burr, C.B. Wenger, E. Gastaldo, and J.W. Gardner. 1996. Exertional heat illness in Marine Corps recruit training. *Aviation, Space, and Environmental Medicine* 67: 354-360.

41. Armstrong, L.E., C.M. Maresh, A.E. Crago, R. Adams, and W.O. Roberts. 1994. Interpretation of aural temperatures during exercise, hyperthermia, and cooling therapy. *Medicine, Exercise, Nutrition and Health* 3: 9-16.

42. Armstrong, L.E., A.E. Crago, R. Adams, W.O. Roberts, and C.M. Maresh. 1996. Whole-body cooling of hyperthermic runners: Comparison of two field therapies. *American Journal of Emergency Medicine* 14: 355-358.

43. Mager, M., R.W. Hubbard, and M.D. Kerstein. 1980. *Survey and analysis of the medical experience for CAX 8-80.* Natick, MA: U.S. Army Research Institute of Environmental Medicine.

44. Al-Marzoogi, A., M. Khagoli, and A. El-Ergesus. 1983. Organizational set-up: Detection, screening, treatment, and follow-up of heat disorders. In *Heat stroke and temperature regulation,* edited by M. Khogali and J.R.S. Hales, 31-40. Sydney, Australia: Academic Press.

45. Lyle, D.M., P.R. Lewis, D.A. Richards, R. Richards, A.E. Bauman, J.R. Sutton, and I.D. Cameron. 1994. Heat exhaustion in the *Sun Herald* City-to-Surf fun run. *Medical Journal of Australia* 161: 361-365.

46. Leithead, C.S., and E.R. Gunn. 1964. The aetiology of cane cutter's cramps in British Guiana. In *Environmental physiology and psychology in arid conditions.* Leige, Belgium: UNESCO.

47. Noakes, T.D., R.J. Norman, R.H. Buck, J. Godlonton, K. Stevenson, and D. Pittaway. 1990. The incidence of hyponatremia during prolonged ultraendurance exercise. *Medicine and Science in Sports and Exercise* 22: 165-170.

48. Speedy, D.B., T.D. Noakes, I.R. Rogers, J.M. Thompson, R.G.D. Campbell, J.A. Kuttner, D.R. Boswell, S. Wright, and M. Hamlin. 1999. Hyponatremia in ultradistance triathletes. *Medicine and Science in Sports and Exercise* 22: 165-170.

49. Speedy, D.B., T.D. Noakes, N.E. Kimber, I.R. Rogers, J.M. Thompson, D.B. Boswell, J.J. Ross, R. Campbell, D.B. Gallagher, and J.A. Kuttner. 2001. Fluid balance during and after an ironman triathlon. *Clinical Journal of Sports Medicine* 11: 44-50.

50. Mulvihill, K. 2001. Runners beware: Too much water can be dangerous. *Health Reuters* [Online], November 30. New York: Reuters America News Service. Available: www.reutershealth.com [4/4/2002].

51. Whayne, T.F. 1951. *History of heat trauma as a war experience.* Washington, DC: Medical Service Officer Basic Course.

52. Terrill, A.A. 1979. *Troop effectiveness and acclimatization in the Middle East and Sub-Sahara Africa.* MacDill Air Force Base, Tampa Bay, FL: Office of the Command Surgeon.

Chapter 3

1. Casa, D.J. 1999. Exercise in the heat I: Fundamentals of thermal physiology, performance implications, and dehydration. *Journal of Athletic Training* 34: 246-252.

2. Rowell, L.B. 1986. *Human circulation regulation during physiological stress.* New York: Oxford University Press.

3. Casa, D.J. 1999. Exercise in the heat II: Critical concepts in rehydration, exertional heat illnesses, and maximizing athletic performance. *Journal of Athletic Training* 34: 253-262.

4. Khogali, M. 1983. Heat stroke: An overview. In *Heat stroke and temperature regulation,* edited by M. Khogali and J.R.S. Hales, 1-12. New York: Academic Press.

5. Dickinson, J.G. 1996. History and epidemiology: Definitions and groups at risk. In *Hyperthermic and hypermetabolic disorders,* edited by P.M. Hopkins and F.R. Ellis, 3-19. New York: Cambridge University Press.

6. Shapiro, Y., and D.S. Moran. 1999. Heat stroke: A consequence of mal-adaptation to heat-exercise exposure. In *Adaptation biology and medicine.* Vol. 2, edited by K.B. Pandolf, N. Takeda, and P.K. Singal, 344-351. New Delhi, India: Narosa Publishing House.

7. Hubbard, R.W. 1990. An introduction: The role of exercise in the etiology of exertional heatstroke. *Medicine and Science in Sports and Exercise* 22: 2-5.

8. Jessen, C. 1987. Thermoregulatory mechanisms in severe heat stress and exercise. In *Heat stress: Physical exertion and environment,* edited by J.R.S. Hales and D.A.B. Richards, 1-20. New York: Excerpta Medica.

9. Hales, J.R.S. 1987. Proposed mechanisms underlying heat stroke. In *Heat stress: Physical exertion and environment,* edited by J.R.S. Hales and D.A.B. Richards, 85-102. New York: Excerpta Medica.

10. Shapiro, Y. 1987. Pathophysiology of hyperthermia and heat intolerance. In *Heat stress: Physical exertion and environment,* edited by J.R.S. Hales and D.A.B. Richards, 263-276. New York: Excerpta Medica.

11. Dickinson, J.G. 1996. Predisposing factors, clinical features, treatment, and prevention. In *Hyperthermic and hypermetabolic disorders,* edited by P.M. Hopkins and F.R. Ellis, 20-41. New York: Cambridge University Press.

12. Knochel, J.P. 1996. Pathophysiology of heat stroke. In *Hyperthermic and hypermetabolic disorders,* edited by P.M. Hopkins and F.R. Ellis, 42-62. New York: Cambridge University Press.

13. Epstein, Y. 2000. Exertional heatstroke: Lessons we tend to forget. *American Journal of Medicine and Sports* 2: 143-152.

14. Shapiro, Y., and D.S. Seidman. 1990. Field and clinical observations of exertional heat stroke patients. *Medicine and Science in Sports and Exercise* 22: 6-14.

15. Sutton, J.R. 1990. Clinical implications of fluid imbalance. In *Fluid homeostasis during exercise,* edited by C.V. Gisolfi and D.R. Lamb, 425-444. Carmel, IN: Brown and Benchmark.

16. Armstrong, L.E., and C.M. Maresh. 1993. The exertional heat illnesses: A risk of athletic participation. *Medicine, Exercise, Nutrition and Health* 2: 125-134.

17. Hubbard, R.W., and L.E. Armstrong. 1989. Hyperthermia: New thoughts on an old problem. *The Physician and Sportsmedicine* 17 (6): 97-113.

18. Sherman, C. 1994. Stress: How to help patients cope. *The Physician and Sportsmedicine* 22: 66-72.

19. Bosenberg, A.T., J.G. Brock-Utne, M.T. Wells, G.T. Blake, and S.L. Gaffin. 1988. Strenuous exercise causes systemic endotoxemia. *Journal of Applied Physiology* 65: 106-108.

20. Brock-Utne, J.G., S.L. Gaffin, M.T. Wells, P. Gathiram, E. Sohor, M.F. James, et al. 1988. Endotoxemia in exhausted runners following a long distance race. *S A Med J* 73: 533-536.

21. Gaffin, S.L., M. Koratich, B. Gentile, N. Leva, and R.W. Hubbard. 1998. A miniswine model of heatstroke. *Journal of Thermal Biology* 23: 341-352.

22. Gaffin, S.L., and R.W. Hubbard. 1996. Experimental approaches to therapy and prophylaxis for heat stress and heatstroke. *Wilderness and Environmental Medicine* 7(4): 312-324.

23. Parillo, J.E., M.M. Parker, C. Natanson, A.F. Suffredini, R.L. Danner, R.E. Cunnion, and F.P. Ognibene. 1990. Septic shock in humans. Advances in the understanding of pathogenesis, cardiovascular dysfunction, and therapy. *Annals of Internal Medicine* 113: 227-242.

24. Hubbard, R.W., and L.E. Armstrong. 1988. The heat illnesses: Biochemical, ultrastructural, and fluid-electrolyte considerations. In *Human performance physiology and environmental medicine at terrestrial extremes,* edited by K.B. Pandolf, M.N. Sawka, and R.R. Gonzalez, 305-359. Indianapolis: Benchmark Press.

25. Hubbard, R.W., S.L. Gaffin, and D.L. Squire. 1995. Heat-related illnesses. In *Management of wilderness and environmental emergencies.* 3d ed., Vol 8, edited by P.S. Auerbach, 167-211. St. Louis: Mosby Year Book.

26. Bouchama, A., R.S. Parhar, A. Er-Yazigi, K. Sheth, and S. Al-Sedairy. 1991. Endotoxemia and release of tumor necrosis factor and interleukin-1-alpha in acute heatstroke. *Journal of Applied Physiology* 70: 2640-2644.

27. Bouchama, A., S. Al-Sedairy, S. Siddiqui, E. Shail, and M. Rezeig. 1993. Elevated pyrogenic cytokines in heatstroke. *Chest* 104: 1498-1502.

28. Bannister, R.G. 1960. Anhidrosis following intravenous bacterial pyrogen. *Lancet* ii: 118-122.

29. Altura, B.M. 1980. Reticuloendothelial cells and host defense. *Advances in Microcirculation* 9: 252-254.

30. Rowell, L.B., G.L. Bregelmann, J.R. Blackman, R.D. Twiss, and F. Kusumi. 1968. Splanchnic blood flow and metabolism in heat-stressed man. *Journal of Applied Physiology* 24: 475-484.

31. Rowell, L.B., J.R. Blackmon, R.H. Martin, J.A. Mazzarella, and R.A. Bruce. 1965. Hepatic clearance of indocyanine green in man under thermal and exercise stress. *Journal of Applied Physiology* 20: 384-394.

32. Rowell, L.B., J. Marx, R.A. Bruce, R.D. Conn, and F. Kusumi. 1966. Reductions in cardiac output, central blood volume and stroke volume with thermal stress in normal men during exercise. *Journal of Clinical Investigation* 45: 1801-1816.

33. Shapiro, Y., M. Alkan, Y. Epstein, F. Newman, and A. Magazanik. 1986. Increase in rat intestinal permeability to endotoxin during hyperthermia. *European Journal of Applied Physiology* 55: 410-412.

34. Gathiram, P., M.T. Wells, J.G. Brock-Utne, and S.L. Gaffin. 1988. Portal and systemic arterial plasma lipopolysaccharide concentrations in heat stressed primates. *Circulatory Shock* 25: 223-230.

35. Sinert, R., L. Kohl, T. Rainone, and T. Scalea. 1994. Exercise-induced rhabdomyolysis. *Annals of Emergency Medicine* 23: 1301-1306.

36. Moeller, J.L., and D.B. McKeag. 1995. Exercise-induced rhabdomyolysis. *Sports Medicine and Arthroscopy Review* 3: 274-279.

37. Vertel, R.M., and J.P. Knochel 1967. Acute renal failure due to heat injury. *American Journal of Medicine* 43: 435-445.

38. Moeller, J.L., and D.B. McKeag. 1995. Exercise-induced rhabdomyolysis. *Sports Medicine and Arthroscopy Review* 3: 274-279.

39. Schrimpf, M., W.S. Queale, and E.G. McFarland. 1999. Exercise-induced rhabdomyolysis in a woman. *Clinical Journal of Sport Medicine* 9: 233-235.

40. Sinert, R., L. Kohl, R. Rainone, and R. Scalea. 1994. Exercise-induced rhabdomyolysis. *Annals of Emergency Medicine* 23 (6): 1301-1306.

41. Armstrong, L.E., J.P. DeLuca, and R.W. Hubbard. 1990. Time course of recovery and heat acclimation ability of prior exertional heatstroke patients. *Medicine and Science in Sports and Exercise* 22 (1): 36-48.

42. Shibolet, S., R. Coll, T. Gilat, and E. Sohar. 1967. Heatstroke: Its clinical picture and mechanism in 36 cases. *Quarterly Journal of Medicine* 36: 525-547.

43. Knochel, J.P. 1996. Clinical complications of body fluid and electrolyte imbalance. In *Body fluid balance: Exercise and sport,* edited by E.R. Buskirk and S.M. Puhl, 297-317. New York: CRC Press.

44. Armstrong, L.E., and C.M. Maresh. 1999. Can humans avoid and recover from exertional heatstroke? In *Adaptation biology and medicine.* Vol. 2, edited by K.B. Pandolf, N. Takeda, and P.K. Singal, 344-351. New Delhi, India: Narosa Publishing House.

45. Armstrong, L.E., C.M. Maresh, A.E. Crago, R. Adams, and W.O. Roberts, 1994. Interpretation of aural temperatures during exercise, hypothermia, and cooling therapy. *Medicine, Exercise, Nutrition and Health* 3: 9-16.

46. Epstein, Y., E. Sohar, and Y. Shapiro. 1995. Exertional heatstroke: A preventable condition. *Israel Journal of Medicine and Science* 31: 454-462.

47. Armstrong, L.E., Y. Epstein, J.E. Greenleaf, E.M. Haymes, R.W. Hubbard, W.O. Roberts, and P.D. Thompson. 1996. American College of Sports Medicine position stand: Heat and cold illnesses during distance running. *Medicine and Science in Sports and Exercise* 28 (12): i-x.

48. Armstrong, L.E. 2000. *Performing in extreme environments.* Champaign, IL: Human Kinetics.

49. Noakes, T.D. 1986. Body cooling as a method for reducing hyperthermia. *South African Medical Journal* 70: 373-374.

50. Adolph, E.F. 1947. *Physiology of man in the desert.* New York: Interscience.

51. Hubbard, R.W., and L.E. Armstrong. 1988. The heat illnesses: Biochemical, ultrastructural, and fluid-electrolyte considerations. In *Human performance physiology and environmental medicine at terrestrial extremes,* edited by K.B. Pandolf, M.N. Sawka, and R.R. Gonzalez, 305-359. Indianapolis: Benchmark Press..

52. Costrini, A. 1990. Emergency treatment of exertional heatstroke and comparison of whole-body cooling techniques. *Medicine and Science in Sports and Exercise* 22: 15-18.

53. Clements, J.M., D.J. Casa, J.C. Knight, J.M. McClung, A.S. Blake, P.M. Meenen, A.M. Gilmer, and K.A. Caldwell. 2002. Ice-water immersion and cold-water immersion provide similar cooling rates in runners with exercise-induced hyperthermia. *Journal of Athletic Training* 37 (2): 146-150.

54. Knight, J.C., D.J. Casa, J.M. McClung, K.A. Caldwell, A.M. Gilmer, P.M. Meenen, and P.J. Goss. 2000. Assessing if two tympanic temperature instruments are valid indicators of core temperature in hyperthermic runners and does drying the ear canal help. *Journal of Athletic Training* 35 (2): S21.

55. Minard, C. 1961. Prevention of heat casualties in Marine Corps recruits. *Military Medicine* 126: 261-272.

56. Noakes, T.D. 1998. Fluid and electrolyte disturbances in heat illness. *International Journal of Sports Medicine* 19: S146-S149.

57. Brodeur, V.B., S.R. Dennett, and L.S. Griffin. 1989. Hyperthermia, ice baths, and emergency care at the Falmouth road race. *Journal of Emergency Nursing* 15 (4): 304-312.

58. Kark, J.A., P.Q. Burr, C.B. Wenger, E. Gastaldo, and J.W. Gardner. 1996. Exertional heat illness in Marine Corps recruit training. *Aviation, Space, and Environmental Medicine* 67 (4): 354-360.

59. Wyndham, C.H., N.B. Strydom, H.M. Cooke, J.S. Maritz, J.F. Morrison, P.W. Fleming, and J.S. Ward. 1959. Methods of cooling subjects with hyperpyrexia. *Journal of Applied Physiology* 14 (3): 771-776.

60. Kielblock, A.J. 1987. Strategies for the prevention of heat disorders with particular reference to the efficacy of body cooling procedures. In *Heat stress: Physical exertion and environment,* edited by J.R.S. Hales and D.A.B. Richards, 489-497. New York: Excerpta Medica.

61. Armstrong, L.E., J.P. DeLuca, R.W. Hubbard, and E.L. Christensen. 1989. *Exertional heatstroke in soldiers: An analysis of predisposing factors, recovery rates, and residual heat intolerance (Report #T5-90).* Natick, MA: United States Army Medical Research and Development Command.

62. Casa, D.J., L.E. Armstrong, S.K. Hillman, S.J. Montain, R.V. Reiff, B.S.E. Rich, W.O. Roberts, and J.A. Stone. 2000. NATA position statement: Fluid replacement for athletes. *Journal of Athletic Training* 35: 212-224.

63. Murray, R., and K. Walsh. 1999. Environmental injuries. In *Athletic training and sports medicine,* edited by R.C. Schenck, 631-652. Rosemont, IL: American Academy of Orthopaedic Surgeons.

64. Armstrong, L.E., C.M. Maresh, S.A. Kavouras, D.J. Casa, J.A. Herrera-Soto,

and F.T. Hacker. 1998. Urinary indices during dehydration, exercise, and rehydration. *International Journal of Sport Nutrition and Exercise Metabolism* 8: 345-355.

65. Murray, R. 2001. Regulation of fluid balance and temperature during exercise in the heat: Scientific and practical considerations. In *Exercise, nutrition, and environmental stress*. Vol. 1, edited by H. Nose, C.V. Gisolfi, and K. Imaizumi, 1-18. Traverse City, MI: Cooper Publishing Group.

66. Morimoto, T., T. Itoh, and A. Takamata. 1998. Thermoregulation and body fluid in hot environment. *Progress in Brain Research* 115: 499-508.

67. Armstrong, L.E., J.P. DeLuca, and R.W. Hubbard. 1990. Time course of recovery and heat acclimation ability of prior exertional heatstroke patients. *Medicine and Science in Sports and Exercise* 22: 36-48.

68. Convertino, V.A., L.E. Armstrong, E.F. Coyle, G.W. Mack, M.N. Sawka, L.C. Senay, and W.M. Sherman. 1996. American College of Sports Medicine position stand: Exercise and fluid replacement. *Medicine and Science in Sports and Exercise* 28 (1): i-vii.

69. Sawka, M.N., and E.F. Coyle. 1999. Influence of body water and blood volume on thermoregulation and exercise performance in the heat. *Exercise and Sport Science Reviews* 27: 167-218.

70. Binkley, H.M., J. Beckett, D.J. Casa, D. Kleiner, and P. Plummer. 2002. National Athletic Trainers Association position statement: Exertional heat illnesses. *Journal of Athletic Training*. 37(3): 329-343.

71. Sawka, M.N., V.A. Convertino, E.R. Eichner, S.M. Schneider, and A.J. Young. 2000. Blood volume: Importance and adaptations to exercise training, environmental stresses, and trauma/sickness. *Medicine and Science in Sports and Exercise* 32 (2): 332-348.

72. Department of the Army. 1995. *Army regulation 40-501: Standards of medical fitness*. Washington, DC: Army Headquarters, pp. 13, 22.

73. Royburt, M., Y. Epstein, Z. Solomon, and J. Shemer. 1993. Long-term psychological and physiological effects of heat stroke. *Physiology and Behavior* 54: 265-267.

74. Mehta, A.C., and R.N. Baker. 1970. Persistent neurological deficits in heat stroke. *Neurology* 20: 336-340.

Chapter 4

1. Weiner, J.S., and G.O. Horne. 1958. A classification of heat illness. *British Medical Journal* 28 (1): 1533-1535.

2. Allen, S.D., and R.P. O'Brien. 1944. Tropical anhidrotic asthenia: Preliminary report. *Medical Journal of Australia* 2: 335-337.

3. Black, D.A., R.A. McCance, and W.F. Young. 1944. A study of dehydration by means of balance experiments. *Journal of Physiology (London)* 102: 406-414.

4. Weiner, J.S. 1938. An experimental study of heat collapse. *Journal of Industrial Hygiene* 20: 389-393.

5. Adolph, E.F. 1947. Water metabolism. *Annual Reviews of Physiology* 9: 381-408.

6. Novy, F.G., and J.H. Ramsey. 1944. Failure of sweat mechanism in the desert. *Journal of the American Medical Association* 125: 738-740.

7. Sulzberger, M.B., H.M. Zimmerman, and K. Emerson. 1946. Tropical an-

hidrotic asthenia and its relationship to prickly heat. *Journal of Investigative Dermatology* 7: 153-155.

8. Wolkin, J., J.J. Goodman, and W.E. Kelley. 1944. Failure of sweat mechanism in desert: Thermogenic anhidrosis. *Journal of the American Medical Association* 124: 479-481.

9. Horne, G.O., and R.H. Mole. 1950. Anhidrotic heat exhaustion. *Transactions of the Royal Society of Tropical Medicine and Hygiene* 44: 193-196.

10. Ladell, W.S.S., J.C. Waterlow, and M.F. Hudson. 1944. Desert climate: Physiological and clinical observations. *Lancet* 2: 491-497.

11. Ladell, W.S.S., J.C. Waterlow, and M.F. Hudson. 1944. Desert climate: Physiological and clinical observations. *Lancet* 2: 527-531.

12. Leithead, C.S., and A.R. Lind. 1964. Heat syncope. In *Heat stress and heat disorders*, edited by C.S. Leithead and A.R. Lind, 136-140. Philadelphia: F.A. Davis Co.

13. Knochel, J.P., and G. Reed. 1987. Disorders of heat regulation. In *Clinical disorders of fluid and electrolyte metabolism*, edited by M.H. Kleeman, C.R. Maxwell, and R.G. Narin, 1197-1232. New York: McGraw-Hill.

14. Yarbrough, B.E., and R.W. Hubbard. 1989. Heat-related illnesses. In *Management of wilderness and environmental emergencies*. 2d ed., edited by P.S. Auerbach and E.C. Geehr, 119-143. St. Louis: C.V. Mosby.

15. Rothstein, A., E.F. Adolph, and J.H. Wills. 1947. Voluntary dehydration. In *Physiology of man in the desert*, edited by E.F. Adolph, 254-270. New York: Interscience.

16. National Center for Health Statistics. 1998. International Classification of Diseases, Ninth Revision, Clinical Modification (ICD-9-CM). Hyattsville, MD: Author.

17. Costrini, A.M., H.A. Pitt, A.B. Gustafson, and D.E. Uddin. 1979. Cardiovascular and metabolic manifestations of heat stroke and severe heat exhaustion. *American Journal of Medicine* 66: 296-302.

18. Petersdorf, R.G. 1994. Hypothermia and hyperthermia. In *Harrison's principles of internal medicine*, edited by K.J. Isselbacher, E. Braunwald, J.D. Wilson, J.B. Martin, A.S. Fauci, and D.L. Kasper, 2473-2479. New York: McGraw-Hill.

19. Armstrong, L.E. 1994. Considerations for replacement beverages: Fluid-electrolyte balance and heat illness. In *Fluid replacement and heat stress*, edited by B.M. Marriott, 37-54. Washington, DC: National Academy Press.

20. Al-D'bbag, M., M. Khogali, and M.Ghallab. 1983. Clinical picture and management of heat exhaustion. In *Heat stroke and temperature regulation*, edited by M. Khogali and J.R.S. Hales, 171-180. Sydney, Australia: Academic Press.

21. Stewart, J.M. 1982. Practical aspects of human heat stress. In *Environmental engineering in South African mines*, edited by J. Burrows, R. Hemp, W. Holding, and R.M. Stroh, 535-567. Yeoville, South Africa: Mine Ventilation Society.

22. Hansen, R.D., T.S. Olds, D.A. Richards, C.R. Richards, and B. Leelarthaepin. 1996. Infrared thermometry in the diagnosis and treatment of heat exhaustion. *International Journal of Sports Medicine* 17: 66-70.

23. Richards, R., D. Richards, and J. Sutton. 1992. Exertion-induced heat

exhaustion: An often overlooked diagnosis. *Australian Family Physician* 21 (1): 18-24.

24. Lyle, D.M., P.R. Lewis, D.A.B. Richards, R. Richards, A.E. Bauman, J.R. Sutton, and I.D. Cameron. 1994. Heat exhaustion in the Sun-Herald City to Surf fun run. *Medical Journal of Australia* 161: 361-365.

25. Bloch, C. 1982. Heat exhaustion during a mountain climb. *South African Medical Journal* 61: 342. [Letter.]

26. Miller, C.W. 1982. Heat exhaustion in tractor drivers. *Occupational Health* 34 (8): 361-365.

27. Hubbard, R.W., and L.E. Armstrong. 1988. The heat illnesses: Biochemical, ultrastructural, and fluid-electrolyte considerations. In *Human performance physiology and environmental medicine at terrestrial extremes,* edited by K.B. Pandolf, M.N. Sawka, and R.R. Gonzalez, 305-359. Indianapolis: Benchmark Press.

28. Armstrong, L.E., Y.E. Epstein, J.E. Greenleaf, E.M. Haymes, R.W. Hubbard, W.O. Roberts, and P.D. Thompson. 1996. American College of Sports Medicine position stand: Heat and cold illnesses during distance running. *Medicine and Science in Sports and Exercise* 28 (12): i-x.

29. Marriott, H.L. 1950. *Water and salt-depletion,* 1-81. Springfield, IL: Charles C. Thomas.

30. Roberts, W.O. 1989. Exercise-associated collapse in endurance events: A classification system. *The Physician and Sportsmedicine* 17: 49-55.

31. Mayers, L.B., and T.D. Noakes. 2000. A guide to treating Ironman triathletes at the finish line. *The Physician and Sportsmedicine* 28 (8): 35-50.

32. Buntman, A.J. 2000. Endurance sports present an assessment dilemma for EMS personnel. *Journal of Emergency Medicine Services* 25 (2): 46-52.

33. Kark, J.A., P.Q. Burr, C.B. Wenger, E. Gastaldo, and J.W. Gardner. 1996. Exertional heat illness in Marine Corps recruit training. *Aviation, Space, and Environmental Medicine* 67: 354-360.

34. Mager, M., R.W. Hubbard, and M.D. Kerstein. 1980. *Survey and analysis of the medical experience for CAX 8-80.* Natick, MA: U.S. Army Research Institute of Environmental Medicine.

35. Al-Marzoogi, A., M. Khogali, and A. El-Ergesus. 1983. Organizational set-up: Detection, screening, treatment, and follow-up of heat disorders. In *Heat stroke and temperature regulation,* edited by M. Khogali and J.R.S. Hales, 31-40. Sydney, Australia: Academic Press.

36. Knochel, J.P. 1974. Environmental heat illness. *Archives of Internal Medicine* 133: 841-864.

37. Armstrong, L.E., J.P. De Luca, and R.W. Hubbard. 1990. Time course of recovery and heat acclimation ability of prior exertional heatstroke patients. *Medicine and Science in Sports and Exercise* 22: 36-48.

38. Hubbard, R.W., S.L. Gaffin, and D. Squire. 1995. Heat-related illnesses. In *Management of wilderness and environmental emergencies.* 3d ed., edited by P.S. Auerbach, 167-211. St. Louis: Mosby Year Book.

39. Armstrong, L.E., R.W. Hubbard, P.C. Szlyk, I.V. Sils, and W.J. Kraemer. 1988. Heat intolerance, heat exhaustion monitored: A case report. *Aviation, Space, and Environmental Medicine* 59: 262-266.

40. McCance, R.A. 1936. Medical problems in mineral metabolism III: Experimental human salt deficiency. *Lancet* 1: 823-834.

41. McCance, R.A. 1936. Medical problems in mineral metabolism III: Experimental sodium chloride deficiency in man. *Proceedings of the Royal Society of London* 119: 245-253.

42. Headquarters, Departments of the Army, Navy, and Air Force. 1980. *Prevention, treatment, and control of heat injury.* Publication TBMED507. Washington, DC: Author.

43. U.S. Department of Health and Human Services. 1986. *Working in hot environments.* DHHS Publication No. 86-112. Washington, DC: Public Health Service, Centers for Disease Control, National Institute for Occupational Safety and Health.

44. Leithead, C.S., and A.R. Lind. 1964. Disorders of water and electrolyte balance. In *Heat stress and heat disorders,* edited by C.S. Leithead and A.R. Lind, 141-177. Philadelphia: F.A. Davis Co.

45. Vaamonde, C.A. 1982. Sodium depletion. In *Sodium: Its biological significance,* edited by S. Papper, 208-234. Boca Raton, FL: CRC Press.

46. Dinman, B.D., and S.M. Horvath. 1984. Heat disorders in industry: A reevaluation of diagnostic criteria. *Journal of Occupational Medicine* 26: 489-495.

47. Shahid, M.S., L. Hatle, H. Mansour, and L. Mimish. 1999. Echocardiographic and doppler study of patients with heatstroke and heat exhaustion. *International Journal of Cardiac Imaging* 15: 279-285.

48. Donoghue, A.M., and G.P. Bates. 2000. The risk of heat exhaustion at a deep underground metalliferous mine in relation to body-mass index and predicted VO$_2$max. *Occupational Medicine* 50: 259-263.

49. Donoghue, A.M., and G.P. Bates. 2000. The risk of heat exhaustion at a deep underground metalliferous mine in relation to surface temperatures. *Occupational Medicine* 50: 334-336.

50. Donoghue, A.M., M.J. Sinclair, and G.P. Bates. 2000. Heat exhaustion in a deep underground metalliferous mine. *Occupational Environmental Medicine* 57: 165-174.

51. Armstrong, L.E., R.W. Hubbard, W.J. Kraemer, J.P. De Luca, and E.L. Christensen. 1988. Signs and symptoms of heat exhaustion during strenuous exercise. *Annals of Sports Medicine* 3: 182-189.

52. Backer, H.D., and S. Collins. 1999. Use of a handheld, battery-operated chemistry analyzer for evaluation of heat-related symptoms in the backcountry of Grand Canyon National Park: A brief report. *Annals of Emergency Medicine* 33 (4): 418-422.

53. Armstrong, L.E., and C.M. Maresh. 1993. The exertional heat illnesses: A risk of athletic participation. *Medicine, Exercise, Nutrition, and Health* 2: 125-134.

54. Noakes, T.D. 1988. Why marathon runners collapse. *South African Medical Journal* 73: 569-571.

55. Kapoor, W.N. 2000. Primary care: Syncope. *New England Journal of Medicine* 343 (25): 1856-1862.

56. Linzer, M., E.H. Yang, N.A.M. Estes, P. Wang, V.R. Vorperian, and W.N. Kapoor. 1997. Clinical guideline: Diagnosing syncope. Part 1: Value of

history, physical examination, and electrocardiography. *Annals of Internal Medicine* 126 (12): 989-996.

57. Fenton, A.M., S.C. Hammill, R.F. Rea, P.A. Low, W. Shen. 2000. Vasovagal syncope. *Annals of Internal Medicine* 133 (9): 714-725.

58. Holtzhausen, L.M., T.D. Noakes, B. Kroning, M. DeKlerk, M. Roberts, and R. Emsley. 1994. Clinical and biochemical characteristics of collapsed ultramarathon runners. *Medicine and Science in Sports and Exercise* 26: 1095-1101.

59. Khosla, R., and K.K. Guntupalli. 1999. Environmental emergencies: Heat-related illnesses. *Critical Care Clinics* 15 (2): 251-263.

60. Costrini, A. 1990. Emergency treatment of exertional heatstroke and comparison of whole body cooling techniques. *Medicine and Science in Sports and Exercise* 22: 15-18.

61. Clapp, A.J., P.A. Bishop, I. Muir, and J.L. Walker. 2001. Rapid cooling techniques in joggers experiencing heat strain. *Journal of Science and Medicine in Sport* 4 (2): 160-167.

62. Armstrong, L.E. 1998. Can humans avoid and recover from exertional heatstroke? In *Adaptation biology and medicine*. Vol. 2, edited by K.B. Pandolf, N. Takeda, and P.K. Singal, 344-351. Japan: Narosa Publishing House.

63. Armstrong, L.E., A.E. Crago, R. Adams, W.O. Roberts, and C.M. Maresh. 1996. Whole-body cooling of hyperthermic runners: Comparison of two field therapies. *American Journal of Emergency Medicine* 14: 355-358.

64. Speedy, D.B., T.D. Noakes, I.R. Rogers, J. Thompson, R.G. Campbell, J.A. Kuttner, D.R. Boswell, S. Wright, and M. Hamlin. 1999. Hyponatremia in ultradistance triathletes. *Medicine and Science in Sports and Exercise* 31 (6): 809-815.

65. Hsieh, M., R. Roth, D.L. Davis, H. Larrabee, and C.W. Callaway. 2002. Hyponatremia in runners requiring on-site medical treatment at a single marathon. *Medicine and Science in Sports and Exercise* 34 (2): 185-189.

66. Noakes, T.D. 2000. Hyponatremia in distance athletes: Pulling the IV on the "dehydration myth." *The Physician and Sportsmedicine* 28 (9): 71-76.

67. Callaham, M.L. 1979. Emergency management of heat illness. In *Emergency physician series*, 1-23. North Chicago, IL: Abbott Laboratories.

68. Nash, H.L. 1985. Treating thermal injury: Disagreement heats up. *The Physician and Sportsmedicine* 13: 134-144.

69. Castellani, J.W., C.M. Maresh, L.E. Armstrong, R.W. Kenefick, D. Riebe, M. Echegaray, D. Casa, and V.D. Castracane. 1997. Intravenous versus oral rehydration: Effects on subsequent exercise-heat stress. *Journal of Applied Physiology* 82: 799-806.

70. Casa, D.J., C.M. Maresh, L.E. Armstrong, S.A. Kavouras, J.A. Herrera, F.T. Hacker, N.R. Keith, and T.A. Elliott. 2000. Intravenous versus oral rehydration during a brief period: Responses to subsequent exercise in heat. *Medicine and Science in Sports and Exercise* 32: 124-133.

71. Castellani, J.W., C.M. Maresh, L.E. Armstrong, R.W. Kenefick, D. Riebe, M. Echegaray, S. Kavouras, and V.D. Castracane. 1998. Endocrine responses during exercise-heat stress: Effects of prior isotonic and hypotonic intravenous rehydration. *European Journal of Applied Physiology and Occupational Physiology* 77: 242-248.

72. Maresh, C.M., J.A. Herrera-Soto, L.E. Armstrong, D.J. Casa, S.A. Kavouras, F.T. Hacker, T.A. Elliott, J. Stoppani, and T.P. Scheett. 2001. Perceptual responses in the heat after brief intravenous versus oral rehydration. *Medicine and Science in Sports and Exercise* 33: 1039-1041.

73. Casa, D.J., C.M. Maresh, L.E. Armstrong, S.A. Kavouras, J.A. Herrera-Soto, F.T. Hacker, T.P. Scheett, and J. Stoppani. 2000. Intravenous versus oral rehydration during a brief period: Stress hormone responses to subsequent exhaustive exercise in the heat. *International Journal of Sport Nutrition and Exercise Metabolism* 10: 361-374.

74. Kenefick, R.W., C.M. Maresh, L.E. Armstrong, J.W. Castellani, D. Riebe, M.E. Echegaray, and S.A. Kavouras. 2000. Plasma vasopressin and aldosterone responses to oral and intravenous saline rehydration. *Journal of Applied Physiology* 89: 2117-2122.

75. Anderson, R.J., G.R. Hart, W.G. Reed, and J. Knochel. 1982. Early assessment and management—Heat injuries. In *Emergency medicine annual*. Vol. 1, edited by I.B. Wolcott and D.A. Rund, 117-140. Norwalk, CT: Appleton-Century Crofts.

76. Hubbard, R.W., and L.E. Armstrong. 1989. Hyperthermia: New thoughts on an old problem. *The Physician and Sportsmedicine* 17 (6): 97-113.

77. Noakes, T.D. 1995. Dehydration during exercise: What are the real dangers? *Clinical Journal of Sport Medicine* 5: 123-128.

78. Noakes, T.D. 2000. Hyperthermia, hypothermia and problems of hydration. In *Endurance in sport*. Vol. 2, 2d ed., edited by R.J. Shephard and P.O. Astrand, 591-613. Oxford, England: Blackwell Science.

79. Noakes, T.D. 1998. Fluid and electrolyte disturbances in heat illness. *International Journal of Sports Medicine* 19: S146-S149.

80. Noakes, T.D. 1999. Perpetuating ignorance: Intravenous fluid therapy in sport. *British Journal of Sports Medicine* 33 (5): 296-297.

81. Smith, H.R., G.S. Dhatt, W.M.A. Melia, and J.G. Dickinson. 1995. Cystic fibrosis presenting as hyponatremic heat exhaustion. *British Medical Journal* 310: 579-580.

82. Snyder, O. 1996. A 17-year-old woman with history of bloody emesis and heat exhaustion. *Journal of Emergency Nursing* 22: 624-625.

83. Sandell, R.C., M.D. Pascoe, and T.D. Noakes. 1988. Factors associated with collapse during and after ultramarathon footraces: A preliminary study. *The Physician and Sportsmedicine* 16 (9): 86-94.

84. Montain, S.J., J.E. Laird, W.A. Latzka, and M.N. Sawka. 1997. Aldosterone and vasopressin responses in the heat: Hydration level and exercise intensity effects. *Medicine and Science in Sports and Exercise* 29 (5): 661-668.

85. Huebner, B. 1997. Back on his feet. *Boston Globe,* June 16, D2.

86. Armstrong, L.E. 2000. *Performing in extreme environments,* 15-70. Champaign, IL: Human Kinetics Publishers.

87. Hine, T. 2000. Hamilton collapses in stretch of 1500. *The Hartford Courant,* October 1, C1.

88. Armstrong, L.E., C.M. Maresh, C.V. Gabaree, J.R. Hoffman, S.A. Kavouras, R.W. Kenefick, J.W. Castellani, and L.E. Ahlquist. 1997. Thermal and circulatory responses during exercise: Effects of hypohydration, dehydration, and water intake. *Journal of Applied Physiology* 82: 2028-2035.

89. Sawka, M.N., A.J. Young, R.P. Francesconi, S.R. Muza, and K.B. Pandolf. 1985. Thermoregulatory and blood responses during exercise at graded hypohydration levels. *Journal of Applied Physiology* 59: 1394-1401.

90. Gonzalez-Alonso, J., R. Mora-Rodriguez, P.R. Below, and E.F. Coyle. 1997. Dehydration markedly impairs cardiovascular function in hyperthermic endurance athletes during exercise. *Journal of Applied Physiology* 82: 1229-1236.

91. Montain, S.J., and E.F. Coyle. 1992. Influence of graded dehydration on hyperthermia and cardiovascular drift during exercise. *Journal of Applied Physiology* 73: 1340-1350.

92. Armstrong, L.E., D.L. Costill, and W.J. Fink. 1985. Influence of diuretic-induced dehydration on competitive running performance. *Medicine and Science in Sports and Exercise* 17: 456-461.

93. Coyle, E.F., and M. Hamilton. 1990. Fluid replacement during exercise: Effects on physiological homeostasis and performance. In *Perspectives in exercise science and sports medicine*. Vol. 3, *Fluid homeostasis during exercise*, edited by C.V. Gisolfi and D.R. Lamb, 281-306. Carmel, IN: Benchmark Press.

94. Brown, A.H. 1947. Dehydration exhaustion. In *Physiology of man in the desert*, edited by E.F. Adolph, 208-217. New York: Interscience.

95. Adolph, E.F. 1947. Life in deserts. In *Physiology of man in the desert*, edited by E.F. Adolph, 326-341. New York: Interscience.

96. Gold, J. 1960. Development of heat pyrexia. *Journal of the American Medical Association* 173: 1175-1182.

97. Astrand, P.O., and K. Rodahl. 1977. *Textbook of work physiology: Physiological basis of exercise*. 2d ed., 55-128. New York: McGraw-Hill.

98. Radigan, L.R., and S. Robinson. 1949. Effects of environmental heat stress and exercise on renal blood flow and filtration rate. *Journal of Applied Physiology* 2: 185-191.

99. Hubbard, R.W. 1990. An introduction: The role of exercise in the etiology of exertional heatstroke. *Medicine and Science in Sports and Exercise* 22: 2-5.

100. Hubbard, R.W., W.T. Matthew, R.E.L. Criss, C. Kelly, I. Sils, M. Mager, W.D. Bowers, and D. Wolfe. 1978. Role of physical effort in the etiology of rat heatstroke injury and mortality. *Journal of Applied Physiology* 45: 463-468.

101. Nadal, J.W., S. Pedersen, and W.G. Maddock. 1941. A comparison between dehydration from salt loss and from water deprivation. *American Journal of Physiology* 134: 691-703.

102. Elkinton, J.R., T.S. Danowski, and A.W. Winkler. 1946. Hemodynamic changes in salt depletion and in dehydration. *Journal of Clinical Investigation* 25: 120-129.

103. Roberts, W.O. 2000. A 12-yr profile of medical injury and illness for the Twin Cities Marathon. *Medicine and Science in Sports and Exercise* 32 (9): 1549-1555.

104. Holtzhausen, L.M., and T.D. Noakes. 1995. The prevalence and significance of post-exercise (postural) hypotension in ultramarathon runners. *Medicine and Science in Sports and Exercise* 27: 1595-1601.

105. Benditt, D.G., S. Sakaguchi, and J.J. Shultz. 1993. Syncope: Diagnostic considerations and role of tilt table testing. *Cardiology Reviews* 1: 146-156.

106. Guyton, A.C., and J.E. Hall. 1996. *Textbook of medical physiology.* 9th ed. Philadelphia: Saunders.

107. Wang, D., S. Sakaguchi, and M. Babcock. 1997. Exercise-induced vasovagal syncope: Limiting the risks. *The Physician and Sportsmedicine* 25: 64-74.

108. Armstrong, L.E., and C.M. Maresh. 1991. The induction and decay of heat acclimatization in trained athletes. *Sports Medicine (New Zealand)* 12: 302-312.

109. Bean, W.B., and L.W. Eichna. 1943. Performance in relation to environmental temperature. *Federation Proceedings* 2: 144-158.

110. Shvartz, E., N.B. Strydom, and H. Kotze. 1975. Orthostatism and heat acclimation. *Journal of Applied Physiology* 39: 590-595.

111. Henderson, Y., A.W. Oughterson, L.A. Greenberg, and C.P. Searle. 1935. Air movement as a stimulus to the skin, the reflex effects upon muscle tonus and indirectly upon the circulation of blood; also the effects of therapeutic baths. *American Journal of Physiology* 114: 269-271.

112. Nielsen, B., S. Strange, N. Christensen, J. Warberg, and B. Saltin. 1997. Acute and adaptive responses to exercise in a warm, humid environment. *Pflugers Archive* 434: 49-56.

Chapter 5

1. Bergeron, M.F. 1996. Heat cramps during tennis: A case report. *International Journal of Sport Nutrition* 6: 62-68.

2. González-Alonso, J., R. Mora-Rodríguez, P.R. Below, and E.F. Coyle. 1997. Dehydration markedly impairs cardiovascular function in hyperthermic endurance athletes during exercise. *Journal of Applied Physiology* 82: 1229-1236.

3. Hargreaves, M., and M. Febbraio. 1998. Limits to exercise performance in the heat. *International Journal of Sports Medicine* 19: S115-S116.

4. Sawka, M.N. 1992. Physiological consequences of hypohydration: Exercise performance and thermoregulation. *Medicine and Science in Sports and Exercise* 24: 657-670.

5. Sawka, M.N., W.A. Latzka, R.P. Matott, and S.J. Montain. 1998. Hydration effects on temperature regulation. *International Journal of Sports Medicine* 19: S108-S110.

6. American Academy of Pediatrics. 2000. Climatic heat stress and the exercising child and adolescent. *Pediatrics* 106: 158-159.

7. Convertino, V.A., L.E. Armstrong, E.F. Coyle, G.W. Mack, M.N. Sawka, L.C. Senay, and W.M. Sherman. 1996. American College of Sports Medicine position stand: Exercise and fluid replacement. *Medicine and Science in Sports and Exercise* 28 (1): i-vii.

8. Casa, D.J., L.E. Armstrong, S.K. Hillman, S.J. Montain, R.V. Reiff, B.S.E. Rich, W.O. Roberts, and J.A. Stone. 2000. National Athletic Trainers' Association position statement: fluid replacement for athletes. *Journal of Athletic Training* 35: 212-224.

9. Benda, C. 1989. Outwitting muscle cramps—is it possible? *The Physician and Sportsmedicine* 17: 173-178.

10. Eaton, J.M. 1989. Is this really a muscle cramp? *Postgraduate Medicine* 86: 227-232.

11. Levin, S. 1993. Investigating the cause of muscle cramps. *The Physician and Sportsmedicine* 21: 111-113.

12. Liu, L., G. Borowski, and L.I. Rose. 1983. Hypomagnesemia in a tennis player. *The Physician and Sportsmedicine* 11: 79-80.

13. Miles, M.P., and P.M. Clarkson. 1994. Exercise-induced muscle pain, soreness, and cramps. *The Journal of Sports Medicine and Physical Fitness* 34: 203-216.

14. O'Toole, M.L., P.S. Douglas, R.H. Laird, and W.D.B. Hiller. 1995. Fluid and electrolyte status in athletes receiving medical care at an ultradistance triathlon. *Clinical Journal of Sport Medicine* 5: 116-122.

15. O'Toole, M.L., P.S. Douglas, C.M. Lebrun, R.H. Laird, T.K. Miller, G.C. Miller, and W.D.B. Hiller. 1993. Magnesium in the treatment of exertional muscle cramps (abstract). *Medicine and Science in Sports and Exercise* 25: S19.

16. O'Toole, M.L., C.M. Lebrun, R.H. Laird, J. James, B.N. Campaigne, P.S. Douglas, and W.D.B. Hiller. 1994. Magnesium for athletes resistant to usual IV treatment (abstract). *Medicine and Science in Sports and Exercise* 26: S205.

17. Stamford, B. 1993. Muscle cramps: Untying the knots. *The Physician and Sportsmedicine* 21: 115-116.

18. Williamson, S.L., R.W. Johnson, P.G. Hudkins, and S.M. Strate. 1993. Exertion cramps: A prospective study of biochemical and anthropometric variables in bicycle riders. *Cycling Science* 5 (1): 15-20.

19. Ogletree, J.W., J.F. Antognini, and G.A. Gronert. 1996. Postexercise muscle cramping associated with positive malignant hyperthermia contracture testing. *American Journal of Sports Medicine* 24 (1): 49-51.

20. Bentley, S. 1996. Exercise-induced muscle cramp: Proposed mechanisms and management. *Sports Medicine* 21: 409-420.

21. Schwellnus, M.P., E.W. Derman, and T.D.Noakes. 1997. Aetiology of skeletal muscle 'cramps' during exercise: A novel hypothesis. *Journal of Sports Sciences* 15: 277-285.

22. Hiller, W.D.B. 1989. Dehydration and hyponatremia during triathlons. *Medicine and Science in Sports and Exercise* 21: S219-S221.

23. Layzer, R.B. 1994. The origin of muscle fasciculations and cramps. *Muscle and Nerve* 17: 1243-1249.

24. Simchak, A.C., and R.M. Pascuzzi. 1991. Muscle cramps. *Seminars in Neurology* 11: 281-287.

25. Bergeron, M.F. 2000. Sodium: The forgotten nutrient. *Sports Science Exchange* 13 (3): 1-4. (Available from the Gatorade Sports Science Institute®, 617 W. Main St., Barrington, IL 60010, USA.)

26. Nadel, E.R., G.W. Mack, and A. Takamata. 1993. Thermoregulation, exercise, and thirst: Interrelationships in humans. In *Perspectives in exercise science and sports medicine.* Vol. 6, *Exercise, heat, and thermoregulation,* edited by C.V. Gisolfi, D.R. Lamb, and E.R. Nadel, 225-256. Dubuque, IA: Brown and Benchmark.

27. Jansen, P.H.P., E.M.G. Joosten, and H.M. Vingerhoets. 1990. Muscle cramp: Main theories as to aetiology. *European Archives of Psychiatry and Neurological Sciences* 239: 337-342.

28. Armstrong, L.E., R.W. Hubbard, B.H. Jones, and J.T. Daniels. 1986. Preparing Alberto Salazar for the heat of the 1984 Olympic Marathon. *The Physician and Sportsmedicine* 14: 73-81.

29. United States Tennis Association. 2001. *Friend at court: The USTA handbook of*

tennis rules and regulations 2001 edition, edited by B. Barber, B. Barr, J. Cummings, B. Farley, S. Gerdes, R. Kaufman, R. Shea, G. Smith, and R. Van Brunt, 120. White Plains, NY: H.O. Zimman.

30. Armstrong, L.E., W.C. Curtis, R.W. Hubbard, R.P. Francesconi, R. Moore, and E.W. Askew. 1993. Symptomatic hyponatremia during prolonged exercise in heat. *Medicine and Science in Sports and Exercise* 25: 543-549.

31. Barr, S.I., D.L. Costill, and W.J. Fink. 1991. Fluid replacement during prolonged exercise: Effects of water, saline, or no fluid. *Medicine and Science in Sports and Exercise* 23: 811-817.

32. Speedy, D.B., T.D. Noakes, I.R. Rogers, J.M. Thompson, R.G. Campbell, J.A. Kuttner, D.R. Boswell, S. Wright, and M. Hamlin. 1999. Hyponatremia in ultradistance athletes. *Medicine and Science in Sports and Exercise* 31: 809-815.

33. Vrijens, D.M., and N.J. Rehrer. 1999. Sodium-free fluid ingestion decreases plasma sodium during exercise in the heat. *Journal of Applied Physiology* 86: 1847-1851.

34. Mulloy, A.L., and R.J. Caruana. 1995. Hyponatremic emergencies. *Medical Clinics of North America* 79: 155-168.

35. Hubbard, R.W., and L.E. Armstrong. 1988. The heat illnesses: Biochemical, ultrastructural, and fluid-electrolyte considerations. In *Human performance physiology and environmental medicine at terrestrial extremes,* edited by K.B. Pandolf, M.N. Sawka, and R.R. Gonzalez, 305-359. Indianapolis: Benchmark Press.

36. Vaamonde, C.A. 1982. Sodium depletion. In *Sodium: Its biological significance,* edited by S. Papper, 207-234. Boca Raton, FL: CRC Press.

37. McGee, S.R. 1990. Muscle cramps. *Archives of Internal Medicine* 150: 511-518.

38. Gisolfi, C.V. 2000. Is the GI system built for exercise? *News in Physiological Sciences* 15: 114-119.

39. Gisolfi, C.V., R.W. Summers, G.P. Lambert, and T. Xia. 1998. Effect of beverage osmolality on intestinal fluid absorption during exercise. *Journal of Applied Physiology* 85: 1941-1948.

40. Daries, H.A., T.D. Noakes, and S.C. Dennis. 2000. Effect of fluid intake volume on 2-h running performances in a 25°C environment. *Medicine and Science in Sports and Exercise* 32: 1783-1789.

41. Armstrong, L.E., C.M. Maresh, J.W. Castellani, M.F. Bergeron, R.W. Kenefick, K.E. LeGasse, and D. Riebe. 1994. Urinary indices of hydration status. *International Journal of Sport Nutrition* 4: 265-279.

42. Armstrong, L.E., J.A. Herrera Soto, F.T. Hacker, Jr., D.J. Casa, S.A. Kavouras, and C.M. Maresh. 1998. Urinary indices during dehydration, exercise, and rehydration. *International Journal of Sport Nutrition* 8: 345-355.

43. Shirreffs, S.M., and R.J. Maughan. 2000. Rehydration and recovery of fluid balance after exercise. *Exercise and Sport Science Reviews* 28: 27-32.

44. Maughan, R.J., J.B. Leiper, and S.M. Shirreffs. 1997. Factors influencing the restoration of fluid and electrolyte balance after exercise in the heat. *British Journal of Sports Medicine* 31: 175-182.

45. Nose, H., G.W. Mack, X. Shi, and E.R. Nadel. 1988. Role of osmolality and plasma volume during rehydration in humans. *Journal of Applied Physiology* 65: 325-331.

46. Sherman, W.M., M.J. Plyley, R.L. Sharp, P.J. Van Handel, R.M. McAllister,

W.J. Fink, and D.L. Costill. 1982. Muscle glycogen storage and its relationship to water. *International Journal of Sports Medicine* 3: 22-24.

47. Allan, J.R., and C.G. Wilson. 1971. Influence of acclimatization on sweat sodium concentration. *Journal of Applied Physiology* 30: 708-712.

48. Wenger, C.B. 1988. Human heat acclimatization, In *Human performance physiology and environmental medicine at terrestrial extremes,* edited by K.B. Pandolf, M.N. Sawka, and R.R. Gonzalez, 153-197 Indianapolis: Benchmark Press.

Chapter 6

1. Hubbard, R.W., and L.E. Armstrong. 1989. Hyperthermia: New thoughts on an old problem. *The Physician and Sportsmedicine* 17: 97-113.

2. Backer, H.D., E. Shopes, S.L. Collins, and H. Barkan. 1999. Exertional heat illness and hyponatremia in hikers. *American Journal of Emergency Medicine* 17: 532-539.

3. Latzka, W.A., and S.J. Montain. 1999. Water and electrolyte requirements for exercise. *Clinics in Sports Medicine* 18 (3): 513-524.

4. Garigan, T., and D.E. Ristedt. 1999. Death from hyponatremia as a result of acute water intoxication in the Army basic trainee. *Military Medicine* 3: 234-238.

5. Shopes, E.M. 1997. Drowning in the desert: Exercise-induced hyponatremia at the Grand Canyon. *Journal of Emergency Nursing* 23: 586-590.

6. Reynolds, N.C., and H.D. Schumaker. 1998. Complications of fluid overload in heat casualty prevention during field training. *Military Medicine* 11: 789-791.

7. Armstrong, L.E., Y. Epstein, J.E. Greenleaf, E.M. Haymes, R.W. Hubbard, W.O. Roberts, and P.D. Thompson. 1996. American College of Sports Medicine position stand: Heat and cold illnesses during distance running. *Medicine and Science in Sports and Exercise* 28 (12): i-x.

8. Montain, S.J., W.A. Latzka, and M.N. Sawka. 1999. Fluid replacement recommendations for training in hot weather. *Military Medicine* 164 (7): 502-508.

9. Flinn, S.D., and R.J. Sherer. 2000. Seizure after exercise in the heat: Recognizing life-threatening hyponatremia. *The Physician and Sportsmedicine* 28 (9): 61-67.

10. U.S. Army Center for Health Promotion and Preventive Medicine. 2000. Hyponatremia hospitalizations, U.S. Army 1989-1996. *Medical Surveillance Monthly Report* 6: 9-11.

11. Speedy, D.B., T.D. Noakes, I.R. Rogers, J.M. Thompson, R.G.D. Campbell, J.A. Kuttner, D.R. Boswell, S. Wright, and M. Hamlin. 1999. Hyponatremia in ultradistance triathletes. *Medicine and Science in Sports and Exercise* 22: 165-170.

12. Davis, D., A. Marino, G. Vilke, J. Dunford, and J. Videen. 1999. Hyponatremia in marathon runners: Experience with the inaugural Rock 'n' Roll Marathon. *Annals of Emergency Medicine* 34: 540-541.

13. Noakes, T.D., R.J. Norman, R.H. Buck, J. Godlonton, K. Stevenson, and D. Pittaway. 1990. The incidence of hyponatremia during prolonged ultraendurance exercise. *Medicine and Science in Sports and Exercise* 22: 165-170.

14. Davis, D.P., J.S. Videen, A. Marino, G.M. Vilke, J.V. Dunford, S.P. Van Camp, and L.G. Maharam. 2001. Exercise-associated hyponatremia in marathon runners: A two-year experience. *Journal of Emergency Medicine* 21: 47-57.

15. Armstrong, L.E., W.C. Curtis, R.W. Hubbard, R.P. Francesconi, R. Moore, and E.W. Askew. 1993. Symptomatic hyponatremia during prolonged exercise in heat. *Medicine and Science in Sports and Exercise* 25: 543-549.

16. Montain, S.J., M.N. Sawka, and C.B. Wenger. 2001. Hyponatremia associated with exercise: Risk factors and pathogenesis. *Exercise and Sport Sciences Reviews* 29: 113-117.

17. Pugh, L.G.C.E., J.L. Corbett, and R.H. Johnson. 1967. Rectal temperature, weight losses, and sweat rates in marathon running. *Journal of Applied Physiology* 23: 347-352.

18. Fallon, K.E., E. Broad, M.W. Thompson, and P.A. Reull. 1998. Nutritional and fluid intake in a 100-km ultramarathon. *International Journal of Sport Nutrition* 8: 24-35.

19. Speedy, D.B., T.D. Noakes, N.E. Kimber, I.R. Rogers, J.M. Thompson, D.B. Boswell, J.J. Ross, R. Campbell, D.B. Gallagher, and J.A. Kuttner. 2001. Fluid balance during and after an ironman triathlon. *Clinical Journal of Sports Medicine* 11: 44-50.

20. Noakes, T.D., N. Goodwin, B.L. Rayner, T. Branken, and R.K.N. Taylor. 1985. Water intoxication: A possible complication during endurance exercise. *Medicine and Science in Sports and Exercise* 17: 370-375.

21. Noakes, T.D. 1995. Hyponatremia of exercise. *News on Sport Nutrition Insider* 3(4): 1-4.

22. O'Toole, M.L., P.S. Douglas, R.H. Laird, and D.B. Hiller. 1995. Fluid and electrolyte status in athletes receiving medical care at an ultradistance triathlon. *Clinical Journal of Sport Medicine* 5: 116-122.

23. Speedy, D.B., I. Rogers, S. Safih, and B. Foley, 2000. Hyponatremia and seizures in an ultradistance triathlete. *The Journal of Emergency Medicine* 18: 41-44.

24. Irving, R.A., T.D. Noakes, R. Buck, R. Van Zyl Smit, E. Raine, J. Godlonton, and R.J. Norman. 1991. Evaluation of renal function and fluid homeostasis during recovery from exercise-induced hyponatremia. *Journal of Applied Physiology* 70: 342-348.

25. Cheng, J.C., D. Zikos, D.R. Peterson, and K.A. Fesher. 1989. Symptomatic hyponatremia: Pathophysiology and management. *Acute Care* 15: 270-292.

26. Speedy, D.B., T.D. Noakes, I.R. Rogers, I. Hellmans, N.E. Kimber, D.R. Boswell, R. Campbell, and J.A. Kuttner. 2000. A prospective study of exercise-associated hyponatremia in two ultradistance triathletes. *Clinical Journal of Sport Medicine* 10: 136-141.

27. Arieff, A. 1987. Hyponatremia associated with permanent brain damage. *Advances in Internal Medicine* 32: 325-344.

28. Arieff, A. 1984. Central nervous system manifestations of disordered sodium metabolism. *Clinics in Endocrinology and Metabolism* 13 (2): 269-294.

29. Ayus, J.C., and A.I. Arieff. 2000. Noncardiogenic pulmonary edema in marathon runners. *Annals of Internal Medicine* 133 (12): 1011.

30. Frizzell, R.T., G.H. Lang, D.C. Lowance, and R. Lanthan. 1986. Hyponatremia and ultramarathon running. *Journal of the American Medical Association* 255: 772-774.

31. Young, M.F., F. Sciurba, and J. Rinaldo. 1987. Delerium and pulmonary edema after completing a marathon. *American Reviews of Respiratory Disease* 136: 737-739.

32. Hiller, W.D.B. 1989. Dehydration and hyponatremia during triathlons. *Medicine and Science in Sports and Exercise* 21: 5219-5221.

33. O'Toole, M.L. 1988. Prevention and treatment of electrolyte abnormalities. In *Medical coverage of endurance athletic events*, edited by R.H. Laird, 93-96. Columbus, OH: Ross Laboratories.

34. Barr, S.I., and D.L. Costill. 1989. Water: Can the endurance athlete get too much of a good thing? *Journal of the American Dietetic Association* 89: 1629-1632.

35. Nelson, P.B., A.G. Robinson, W. Kapoor, and J. Rinaldo. 1988. Hyponatremia in a marathoner. *The Physician and Sportsmedicine* 16: 78-88.

36. Speedy, D.B., T.D. Noakes, and C. Schneider. 2001. Exercise-associated hyponatremia: A review. *Emergency Medicine* 13: 17-27.

37. Rowell, L.B. 1986. Hepatic metabolism during hyperthermia and hypoxemia. In *Biochemistry of exercise VI*, edited by B. Saltin, 203-205. Champaign, IL: Human Kinetics.

38. Radigan, L.R., and S. Robinson. 1949. Effects of environmental heat stress and exercise on renal blood flow and filtration rate. *Journal of Applied Physiology* 2: 185-191.

39. Milvy, P. 1977. *The marathon: Physiological, medical, epidemiological and psychological studies.* New York: New York Academy of Sciences.

40. Yunusov, S.Y. 1964. Water-salt metabolism in hot climates. In *Environmental physiology and psychology in arid conditions*, 357-361. Leige, Belgium: UNESCO.

41. Noakes, T.D. 1992. The hyponatremia of exercise. *International Journal of Sport Nutrition* 2: 205-228.

42. Gisolfi, C.V., K.J. Spranger, R.W. Summers, H.P. Schedl, and T.L. Bleiler. 1991. *Journal of Applied Physiology* 71: 2518-2527.

43. Smyth, F.S., W.C. Deamer, and N.M. Phatak. 1933. Studies in so-called water intoxication. *Journal of Clinical Investigation* 12: 55-65.

44. Jose, C.J., and J. Perez-Cruet. 1979. Incidence and morbidity of self-induced water intoxication in state mental hospital patients. *American Journal of Psychiatry* 136 (2): 221-222.

45. Kirch, D.G., L.B. Bigelow, D.R. Weinberger, W.B. Lawson, and R.J. Wyatt. 1985. Polydipsia and chronic hyponatremia in schizophrenic inpatients. *Journal of Clinical Psychiatry* 46 (5): 179-181.

46. Barlow, E.D., and H.E. DeWardener. 1959. Compulsive water drinking. *Quarterly Journal of Medicine* 28: 235-258.

47. Goldberger, E. 1980. *A primer of water, electrolyte and acid-base syndromes.* 6th ed., 58-120. Philadelphia: Lea and Febiger.

48. Pollock, A.S., and A.I. Arieff. 1980. Abnormalities of cell volume regulation and their functional consequences. *American Journal of Physiology* 239: F195-F205.

49. Chung, H.M., R.J. Anderson, R. Kluge, and R.W. Schrier. 1984. Hyponatremia: A prospective study of frequency, cause, and outcome. *Kidney International* 25: 160-167.

50. Helwig, F.C., C.B. Schultz, and D.E. Curry. 1935. Water intoxication: Report of a fatal case with clinical, pathological, and experimental studies. *Journal of the American Medical Association* 104: 1569-1574.

51. Rashkind, M. 1974. Psychosis, polydipsia, and water intoxication: Report of a fatal case. *Archives of General Psychiatry* 30: 112-116.

52. Rendell, M., D. McGrane, and M. Cuesta. 1978. Fatal compulsive water drinking. *Journal of the American Medical Association* 240: 2557-2559.

53. Holliday, M.A., M.N. Kalayci, and J. Harrah. 1968. Factors that limit brain volume changes in response to acute and sustained hyper- and hyponatremia. *Journal of Clinical Investigation* 47: 1916-1924.

54. McManus, M.L., K.B. Churchwell, and K. Strange. 1995. Regulation of cell volume in health and disease. *New England Journal of Medicine* 333: 1260-1266.

55. Speedy, D.B., I.R. Rogers, T.D. Noakes, S. Wright, J.M. Thompson, R. Campbell, I. Hellemans, N.E. Kimber, D.R. Boswell, J.A. Kuttner, and S. Safih. 2000. Exercise-induced hyponatremia in ultradistance triathletes is caused by inappropriate fluid retention. *Clinical Journal of Sports Medicine* 10: 272-278.

56. Galun, E., I. Tur-Kaspa, E. Assia, R. Burstein, N. Strauss, Y. Epstein, and M.M. Popovtzer. 1991. Hyponatremia induced by exercise: A 24-hour endurance march study. *Mineral and Electrolyte Metabolism* 17: 315-320.

57. Travenol Laboratories. 1982. *Fundamentals of fluid and electrolyte imbalances*, 4-26. Deerfield, IL: Author.

58. Noakes, T.D. 2000. Hyponatremia in distance athletes: Pulling the IV on the "dehydration myth." *The Physician and Sportsmedicine* 28: 71-76.

59. Rogers, G., C. Goodman, and C. Rosen. 1997. Water budget during ultra-endurance exercise. *Medicine and Science in Sports and Exercise* 29: 1477-1481.

60. Nose, H., G.W. Mack, X. Shi, and E.R. Nadel. 1988. Shift in body fluid compartments after dehydration in man. *Journal of Applied Physiology* 65: 318-324.

61. Romero, J.C., R.J. Stameloni, M.L. Dafu, R. Dohmen, A. Binia, B. Kluman, and J.C. Fasciolo. 1968. Changes in fluid compartments, renal hemodynamics, plasma renin, and aldosterone secretion induced by low sodium intake. *Metabolism* 17: 10-19.

62. Dill, D.B., and D.L. Costill. 1974. Calculation of percentage changes in volumes of blood, plasma, and red cells in dehydration. *Journal of Applied Physiology* 37: 247-248.

63. Rehrer, N.J., F. Brouns, E.J. Beckers, W.O. Frey, B. Villiger, C.J. Riddoch, P. Menheere, and W.H.M. Saris. 1996. Physiological changes and gastrointestinal symptoms as a result of ultra-endurance running. *European Journal of Applied Physiology and Occupational Physiology* 73: 1-8.

64. Pastene, J., M. German, A.M. Allevard, C. Gharib, and J.R. Lacour. 1996. Water balance during and after marathon running. *European Journal of Applied Physiology and Occupational Physiology* 73: 49-55.

65. Pivarnik, J.M., E.M. Leeds, and J.E. Wilkerson. 1984. Effects of endurance exercise on metabolic water production and plasma volume. *Journal of Applied Physiology* 56: 613-618.

66. Speedy, D.B., T.D. Noakes, T. Boswell, J.M.D. Thompson, N. Rehrer, and D.R.

Boswell. 2001. Response to a fluid load in athletes with a history of exercise induced hyponatremia. *Medicine and Science in Sports and Exercise* 33: 1434-1442.

67. Berkeley Scientific Publications. 1984. *Alba's medical technology board examination review*. Vol. 1, 10th ed. Anaheim, CA: Author.

68. Renkin E.M., and P.M. Cala. 1978. Two take-home problem sets for use in teaching elementary fluid and electrolyte physiology to 1st year medical students in an integrated cell biology course. *The Physiologist* 21: 28-30.

69. Gabow, P.A., W.D. Kaehney, and S.P. Kelleher. 1982. The spectrum of rhabdomyolysis. *Medicine* 61: 141-152.

70. Heitzman, E.J., J.F. Patterson, and M.M. Stanley. 1962. Myoglobinuria and hypokalemia in regional enteritis. *Archives of Internal Medicine* 110: 155-162.

71. Greenberg, J., and L. Arneson. 1967. Exertional rhabdomyolysis with myoglobinuria in a large group of military trainees. *Neurology* 17: 216-222.

72. Knochel, J.P., L.N. Dotin, and R.J. Hamburger. 1972. Pathophysiology of intense physical conditioning in a hot climate. I. Mechanisms of potassium depletion. *Journal of Clinical Investigation* 51: 242-249.

73. Knochel, J.P., and E.M. Schlein. 1972. On the mechanism of rhabdomyolysis in potassium depletion. *Journal of Clinical Investigation* 51: 1750-1758.

74. Putterman, C., L. Levy, and D. Rubinger. 1993. Transient exercise-induced water intoxication and rhabdomyolysis. *American Journal of Kidney Diseases* 21: 206-209.

75. Adler, S. 1980. Hyponatremia and rhabdomyolysis: A possible relationship. *Southern Medical Journal* 73 (4): 511-512.

76. Verbalis, J.G., and S.R. Gullans. 1993. Rapid correction of hyponatremia produces differential effects on brain osmolyte and electrolyte reaccumulation in rats. *Brain Research* 606: 19-27.

77. Dawes, G.S. 1941. The vasodilator action of potassium. *Journal of Physiology (London)* 99: 224-230.

78. Kjellmer, I. 1961. Role of K ions in exercise hyperemia. *Medicina Experimentalis* 5: 50-55.

79. Nagara, I.S. 1998. Intravenous versus oral rehydration. *British Journal of Sports Medicine* 32: 265.

80. Clark, J.M., and F.J. Gennari. 1993. Encephalopathy due to severe hyponatremia in an ultramarathon runner. *Western Journal of Medicine* 159: 188-189.

81. Speedy, D.B., I.R. Rogers, T.D. Noakes, J. Guirey, D.R. Boswell, and S. Safih. 2000. Diagnosis and prevention of hyponatremia at an ultradistance triathlon. *Clinical Journal of Sports Medicine* 10: 52-58.

82. Worthley, L.I.G., and P.D. Thomas. 1986. Treatment of hyponatremic seizures with intravenous 29.2% saline. *British Medical Journal* 292: 168-170.

83. Cluitmans, F.H.M., and A.E. Meinders. 1990. Management of severe hyponatremia: Rapid or slow correction? *The American Journal of Medicine* 88: 161-166.

84. Berl, T. 1990. Treating hyponatremia: What is all the controversy about? *Annals of Internal Medicine* 113: 417-419.

85. Sterns, R.H. 1990. The treatment of hyponatremia: First, do no harm. *The American Journal of Medicine* 88: 557-560.

86. Ayus J.C., R.K. Krothapalli, and A.I. Arieff. 1987. Treatment of symptomatic hyponatremia and its relation to brain damage: A prospective study. *New England Journal of Medicine* 317 (19): 1190-1195.

87. Ng, R.C. 1991. Treatment of hyponatremia. *Annals of Internal Medicine* 114: 248.

88. Armstrong, L.E. 1994. Hyponatremia in endurance athletes. *Sports Medicine Digest (UCLA)* 16 (9): 1-5.

89. Gowrishankan, M., C. Ching-Bun, C. Surinder, A. Steele, and M.L. Halperin. 1997. Hyponatremia in the rat in the absence of positive water balance. *Journal of the American Society of Nephrology* 8: 524-529.

90. Convertino, V.A., L.E. Armstrong, E.F. Coyle, G.W. Mack, M.N. Sawka, L.C. Senay, W.M. Sherman. 1996. American College of Sports Medicine position stand: Exercise and fluid replacement. *Medicine and Science in Sports and Exercise* 28: i-vii.

91. Basnyat, B., J. Sleggs, and M. Spinger. 2000. Seizures and delerium in a trekker: The consequences of excessive water drinking? *Wilderness and Environmental Medicine* 11: 69-70.

Chapter 7

1. Allen, S.D., and J.P. O'Brien. 1944. Tropical anhidrotic asthenia: A preliminary report. *The Medical Journal of Australia* 2: 335-336.

2. Wolkin, J., J.I. Goodman, and W.E. Kelley. 1944. Failure of the sweat mechanism in the desert: Thermogenic anhidrosis. *Journal of the American Medical Association* 12: 478-482.

3. Stewart, D. 1945. Therapeutic use of sodium chloride in industry. *British Journal of Industrial Medicine* 2: 102.

4. Horne, G.O., and R.H. Mole. 1951. Mammillaria. *Transactions of the Royal Society of Tropical Medicine and Hygiene* 44 (4): 465-471.

5. Drake, K., and S.M. Nettina. 1994. Recognition and management of heat-related illness. *Nurse Practitioner* 19 (8): 43-47.

6. Yarbrough, B.E., and R.W. Hubbard. 1989. Heat-related illnesses. In *Management of wilderness and environmental emergencies,* edited by P.S. Auerbach and E.C. Geehr, 119-143. St. Louis: CV Mosby.

7. Pandolf, K.B., T.B. Griffin, E.H. Munro, and R.F. Goldman. 1980. Persistence of impaired heat tolerance from artificially induced miliaria rubra. *American Journal of Physiology* 239 (3): R226-R232.

8. Shelley, W.B. 1953. Miliaria. *Journal of the American Medical Association* 152: 670-672.

9. Straka, B.F., P.H. Cooper, and K.E. Greer. 1991. Congenital miliaria crystallina. *Cutis.* 47: 103-106.

10. Arpey, C.J., L.S. Nagashima-Whalen, M.M. Chren, and M.T. Zaim. 1992. Congenital miliaria crystallina: Case report and literature review. *Pediatric Dermatology* 9 (3): 283-287.

11. Feng, E., and C.K. Janniger. 1995. Miliaria. *Dermatology and Pediatrics* 55: 213-216.

12. Kirk, J.F., B.B. Wilson, W. Chun, and P.H. Cooper. 1996. Miliaria profunda. *Journal of the American Academy of Dermatology* 35 (5): 854-856.

13. Rogers, M., A. Kan, K. Stapleton, and A. Kemp. 1990. Giant centrifugal miliaria profunda. *Pediatric Dermatology* 7 (2): 140-146.

14. Sulzberger, M.B., and D.R. Harris. 1972. Miliaria and anhidrosis: Multiple small patches and the effects of different periods of occlusion. *Archives of Dermatology* 105: 845-850.

15. Pandolf, K.B., T.B. Griffin, E.H. Munro, and R.F. Goldman. 1980. Heat intolerance as a function of percent of body surface involved with miliaria rubra. *American Journal of Physiology* 239 (3): R233-R240.

16. Donoghue, A.M., and M.J. Sinclair. 2000. Miliaria rubra of the lower limbs in underground miners. *Occupational Medicine* 50 (6): 430-433.

17. Holzle, E., and A.M. Kligman. 1978. The pathogenesis of miliaria rubra: Role of the resident microflora. *British Journal of Dermatology* 99: 117-137.

18. Leithead, C.S., and A.R. Lind. 1964. *Heat stress and heat disorders*, 178-194. Philadelphia: F.A. Davis Co.

19. Armstrong, L.E., R.W. Hubbard, Y. Epstein, and R. Weien. 1988. Nonconventional remission of miliaria rubra during heat acclimation: A case report. *Military Medicine* 153: 402-404.

20. Pandolf K.B., R.W. Gagne, W.A. Latzka, I.H. Blank, K.K. Kraning, and R.R. Gonzalez. 2000. Human thermoregulatory responses during heat exposure after artificially induced sunburn. *Journal of Applied Physiology* 262 (4): R610-R616.

21. Leffell, D.J. 2000. The other side of the sun. *Lancet* 356 (9231): 699-700.

22. Saraiya, M., H.I. Hall, and R.J. Uhler. 2002. Sunburn prevalence among adults in the United States, 1999. *American Journal of Preventive Medicine* 23: 91-97.

23. Amundson, L.H. 1990. Managing skin problems in athletes. In *The team physician's handbook*, edited by M.B. Mellion, W.M. Walsh, and G.L. Shelton, 236-250. St. Louis: Mosby-Year Book.

24. Gabaree, C.L., B.S. Mair, M.A. Kolka, and L.A. Stephenson. 1997. Effects of topical skin protectant on heat exchange in humans. *Aviation, Space and Environmental Medicine* 68: 1019-1024.

25. Spaul, W.A., J.A. Boatman, S.W. Emling, H.G. Dirks, S.B. Flohr, W.H. Crocker, and M.A. Glazeski. 1985. Reduced tolerance for heat stress environments caused by protective lotions. *American Industrial Hygiene Association Journal* 46: 460-462.

26. Lee-Chiong, Jr., T.L., and J.T. Stitt. 1995. Heatstroke and other heat-related illnesses: The maladies of summer. *Postgraduate Medicine* 98: 26-36.

Chapter 8

1. Armstrong, L.E., J.P. DeLuca, and R.W. Hubbard. 1990. Time course of recovery and heat acclimation ability of prior exertional heatstroke patients. *Medicine and Science in Sports and Exercise* 22: 36-48.

2. Epstein, Y. 1990. Heat intolerance: Predisposing factor or residual injury? *Medicine and Science in Sports and Exercise* 22: 19-35.

3. Shapiro, Y., and D.S. Seidman. 1990. Field and clinical observations of exertional heat stroke patients. *Medicine and Science in Sports and Exercise* 22: 6-14.

4. Yarbrough, B.E, and R.W. Hubbard. 1989. Heat-related illness. In *Management of wilderness and environmental emergencies.* Vol. 2, edited by P.S. Auerbach and E.C. Geehr. St. Louis: CV Mosby.

5. Shibolet, S., M.C. Lancaster, and Y. Danon. 1976. Heat stroke: A review. *Aviation, Space, and Environmental Medicine* 47: 280-301.

6. Bartley, J.D. 1977. Heat stroke: Is total prevention possible? *Military Medicine* 142: 528-535.

7. Knochel, J.P. 1974. Environmental heat illness: An eclectic review. *Archives of Internal Medicine* 133: 841-864.

8. Casa, D.J. 1999. Exercise in the heat. II. Critical concepts in rehydration, exertional heat illness, and maximizing athletic performance. *Journal of Athletic Training* 34: 253-262.

9. Hubbard, R.W., S.L. Gaffin, and D. Squire. 1995. Heat-related illness. In *Management of wilderness and environmental emergencies.* Vol. 8, edited by P.S. Auerbach, 167-211. St. Louis: Mosby-Year Book.

10. Armstrong, L.E., Y. Epstein, J.E. Greenleaf, E.M. Haymes, R.W. Hubbard, W.O. Roberts, and P.D. Thompson. 1996. American College of Sports Medicine position stand: Heat and cold illnesses during distance running. *Medicine and Science in Sports and Exercise* 28 (12): i-x.

11. Kark, J.A., P.Q. Burr, C.B. Wenger, E. Gastaldo, and J.W. Gardner. 1996. Exertional heat illness in Marine Corps recruit training. *Aviation, Space, and Environmental Medicine* 67: 354-360.

12. Payen, J.F., L. Bourdon, H. Reutenauer, B. Melin, J.F. LeBas, P. Stieglitz, and M. Cure. 1992. Exertional heatstroke and muscle metabolism: An *in vivo* ^{31}P-MRS study. *Medicine and Science in Sports and Exercise* 24: 420-425.

13. Al-Khawashki, M.I., M.K.Y. Mustafa, M. Khogali, and H. El-Sayed. 1983. Clinical presentation of 172 heat stroke cases seen at Mina and Arafat, September 1982. In *Heat stroke and temperature regulation*, edited by M. Khogali and J.R.S. Hales, 99-108. Sydney, Australia: Academic Press.

14. Strydom, N.B. 1971. Age as a causal factor in heat stroke. *Journal of the South African Institute of Mining and Metallurgy* 72: 112-114.

15. Mehta, A.C., and R.N. Baker. 1970. Persistent neurological deficits in heatstroke. *Neurology* 20: 336-340.

16. Keren, G., Y. Epstein, and A. Magazanik. 1981. Temporary heat intolerance in a heatstroke patient. *Aviation, Space, and Environmental Medicine* 52: 116-117.

17. Shapiro, Y., A. Magazanik, R. Udassin, G. Ben-Barnch, E. Shvartz, and Y. Shoenfeld. 1979. Heat intolerance in former heatstroke patients. *Annals of Internal Medicine* 90: 913-916.

18. Hubbard, R.W., and L.E. Armstrong. 1989. Hyperthermia: New thoughts on an old problem. *Physician and Sportsmedicine* 17 (6): 97-113.

19. Shapiro, Y. 1987. Pathophysiology of hyperthermia and heat intolerance. In *Transaction of the Menzies Foundation.* Vol. 14, edited by J.R.S. Hales and D.A.B. Richards, 133-142. Melbourne, Australia: Menzies Foundation.

20. Pandolf, K.B., T.B. Griffin, E.H. Munro, and R.F. Goldman. 1980. Persistence of impaired heat tolerance from artificially induced miliaria rubra. *American Journal of Physiology* 239: R226-R232.

21. Kuno, Y. 1956. *Human perspiration.* Springfield, IL: Charles C. Thomas.

22. Armstrong, L.E., R. Hubbard, Y. Epstein, and R. Weien. 1988. Nonconventional remission of miliaria rubra during heat acclimation. *Military Medicine* 153: 402-404.

23. Shibolet, S., R. Coll, and T. Gilat. 1967. Heatstroke: Its clinical picture and mechanisms in 36 cases. *Quarterly Journal of Medicine* 36: 525-548.

24. Lavenne, F., and D. Belayew. 1966. Exercise tolerance test at room temperature for the purpose of selecting rescue teams for training in a hot climate. *Revue de l'Institut d'Hygiene des Mines* 21: 48-58.

25. Shvartz, E., S. Shibolet, A. Merez, A. Magazanik, and Y. Shapiro. 1977. Prediction of heat tolerance from heart rate and rectal temperature in a temperate environment. *Journal of Applied Physiology* 43: 684-688.

26. Fruth, J.M., and C.V. Gisolfi. 1983. Work-heat tolerance in endurance-trained rats. *Journal of Applied Physiology* 54 (1): 249-253.

27. Armstrong, L.E., and K.B. Pandolf. 1988. Physical training, cardiorespiratory physical fitness, and exercise-heat tolerance. In *Human performance physiology and environmental medicine at terrestrial extremes*, edited by K.B. Pandolf, M.N. Sawka, and R.R. Gonzalez, 199-226. Indianapolis: Benchmark Press.

28. Armstrong, L.E., and C.M. Maresh. 1991. The induction and decay of heat acclimatization in trained athletes. *Sports Medicine* 12: 302-312.

29. Armstrong, L.E., and C.M. Maresh. 1998. Effects of training, environment, and host factors on the sweating response to exercise. *International Journal of Sports Medicine* 19: S103-S105.

30. Seppa, N. 2001. Does lack of sleep lead to diabetes? *Science News* 160 (2): 31.

31. Vroman, N.B., E.R. Buskirk, and J.L. Hodgson. 1983. Cardiac output and skin blood flow in lean and obese individuals during exercise in the heat. *Journal of Applied Physiology* 55: 69-74.

32. Bar-Or, O., H. Lundegren, and E.R. Buskirk. 1969. Heat tolerance of exercising obese and lean women. *Journal of Applied Physiology* 26 (4): 403-409.

33. Buskirk, E.R., H. Lundegren, and L. Magnusson. 1965. Heat acclimation patterns in obese and lean individuals. *Annals of the New York Academy of Sciences* 131: 637-653.

34. Kenney, W.L. 1985. Physiological correlates of heat intolerance. *Sports Medicine* 2: 279-286.

35. Lyle, D.M., P.R. Lewis, D.A. Richards, R. Richards, A.E. Bauman, J.R. Sutton, and I.D. Cameron. 1994. Heat exhaustion in the Sun Herald City to Surf fun run. *Medical Journal of Australia* 161: 361-365.

36. England, A.C., D.W. Fraser, A.W. Hightower, R. Tirinnanzi, D.J. Greenberg, K.E. Powell, C.M. Slovis, and R.A. Varsha. 1982. Preventing severe heat injury in runners: Suggestions from the 1979 Peachtree Road Race experience. *Annals of Internal Medicine* 97: 196-201.

37. Khogali, M. 1982. *Heat disorders with special reference to the Makkah Pilgrimage (The Hajj)*. Saudi Arabia: Ministry of Health, Western Region.

38. Shvartz, E., N.B. Strydom, and H. Kotze. 1975. Orthostatism and heat acclimation. *Journal of Applied Physiology* 39 (4): 590-595.

39. Hubbard, R.W., and L.E. Armstrong. 1988. The heat illnesses: Biochemical, ultrastructural and fluid-electrolyte considerations. In *Human performance physiology and environmental medicine at terrestrial extremes*, edited by K.B. Pandolf, 305-360. Indianapolis: Benchmark Press.

40. Hallivil, J.R. 2001. Mechanisms and clinical implications of post-exercise hypotension in humans. *Exercise and Sport Science Reviews* 29: 65-70.

41. Hargarten, K.M. 1992. Syncope. *The Physician and Sportsmedicine* 20 (5): 123-133.

42. Ladell, W.S.S. 1949. Heat cramps. *Lancet* 2: 836-839.

43. Leithhead, C.S., and A.R. Lind. 1964. *Heat stress and heat disorders.* Philadelphia: F.A. Davis Co.

44. Callaham, M.L. 1979. *Emergency management of heat illness.* North Chicago, IL: Abbott Laboratories.

45. Talbott, J.H. 1935. Heat cramps. In *Medicine,* 323-376. Baltimore, MD: Williams and Wilkins.

46. Malhotra, M.S., and Y. Venkataswamy. 1974. Heat casualties in the Indian Armed Forces. *Indian Journal of Medical Research* 62: 1293-1302.

47. Talbott, J.H., and J. Michelsen. 1933. Heat cramps. *Journal of Clinical Investigation* 12: 533-535.

48. Moss, K.N. 1922. Some effects of high air temperatures and muscular exertion upon colliers. *Proceedings of the Royal Society, Series B* 95: 181-183.

49. Haldane, J.S. 1923. Heat cramp. *British Medical Journal* 1: 609.

50. Leithhead, C.S., and E.R. Gunn. 1964. The aetiology of cane cutter's cramps in British Guiana. In *Environmental physiology and psychology in arid conditions,* 13-17. Leige, Belgium: UNESCO.

51. American Medical Association. 1998. *International classification of diseases.* 9th ed. Chicago: Author.

52. Montain, S.J., M.N. Sawka, and C.B. Wenger. 2001. Hyponatremia associated with exercise: Risk factors and pathogenesis. *Exercise and Sport Sciences Reviews* 29: 113-117.

53. Backer, H.D., E. Shopes, S.L. Collins, and H. Barkam. 1999. External heat illness and hyponatremia in hikers. *American Journal of Emergency Medicine* 17: 532-539.

54. Latzka, W.A., and S.J. Montain. 1999. Water and electrolyte requirements for exercise. *Clinics in Sports Medicine* 18 (3): 513-524.

55. Garigan, T., and D.E. Ristedt. 1999. Death from hyponatremia as a result of acute water intoxication in an Army basic trainee. *Military Medicine* 164 (3): 234-238.

56. Shopes, E.M. 1997. Drowning in the desert: Exercise-induced hyponatremia at the Grand Canyon. *Journal of Emergency Nursing* 23: 586-590.

57. Reynolds, N.C., S. Feighery, and H.D. Schumaker. 1998. Complications of fluid overload in heat casualty prevention during field training. *Military Medicine* 163 (11): 789-791.

58. Montain, S.J., W.A. Latzka, and M.N. Sawka. 1999. Fluid replacement recommendations for training in hot weather. *Military Medicine* 164 (7): 502-508.

59. Flinn, S.D., and R.J. Sherer. 2000. Seizure after exercise in the heat: Recognizing life-threatening hyponatremia. *The Physician and Sportsmedicine* 28 (9): 61-67.

60. Armstrong, L.E. 1994. Hyponatremia in endurance athletes. *Sports Medicine Digest (UCLA)* 16 (9): 1-5.

61. Davis, D.P., J.S. Videen, A. Marmo, G.M. Vilke, J.V. Dunford, S.P. Van Camp, and L.G. Maharam. 2001. Exercise-associated hyponatremia in marathon runners: A two-year experience. *Journal of Emergency Medicine* 21: 47-57.

62. Noakes, T.D., R.J. Norman, R.H. Buck, J. Godlonton, K. Stevenson, and D. Pittaway. 1990. The incidence of hyponatremia during prolonged ultraendurance exercise. *Medicine and Science in Sports and Exercise* 22: 165-170.

63. Speedy, D.B., T.D. Noakes, I.R. Rogers, J.M. Thompson, R.G.D. Campbell, J.A. Kuttner, D.R. Boswell, S. Wright, and M. Hamlin. 1999. Hyponatremia in ultradistance triathletes. *Medicine and Science in Sports and Exercise* 31: 809-815.

64. Davis, D., A. Marino, G. Vilke, J. Dunford, and J. Videen. 1999. Hyponatremia in marathon runners: Experience with the inaugural Rock 'n' Roll Marathon. *Annals of Emergency Medicine* 34: 540-541.

65. U.S. Army Center for Health Promotion and Preventive Medicine. 2000. Hyponatremia hospitalizations, U.S. Army 1989-1996. *Medical Surveillance Monthly Report* 6: 9-11.

66. Armstrong, L.E., W.C. Curtis, R.W. Hubbard, R.P. Francesconi, R. Moore, and E.W. Askew. 1993. Symptomatic hyponatremia during prolonged exercise in heat. *Medicine and Science in Sports and Exercise* 25: 543-549.

67. Ayus, J.C., J. Varon, and A.I. Arieff. 2000. Hyponatremia, cerebral edema, and noncardiogenic pulmonary edema in marathon runners. *Annals of Internal Medicine* 132: 711-714.

68. Speedy, D.B., T.D. Noakes, and C. Schneider. 2001. Exercise-associated hyponatremia: A review. *Emergency Medicine* 13: 17-27.

69. Armstrong, L.E., R.W. Hubbard, J.P. DeLuca, and E.L. Christensen. 1987. Heat acclimatization during summer running in the northeastern United States. *Medicine and Science in Sports and Exercise* 19: 131-136.

70. Vrijens, D.M.D., and N.J. Rehrer. 1999. Sodium-free fluid ingestion decreases plasma sodium during exercise in the heat. *Journal of Applied Physiology* 86: 1847-1851.

71. Smith H.R., G.S. Dhatt, W.M.A. Melia, and J.G. Dickinson. 1995. Lesson of the week: Cystic fibrosis presenting as hyponatremic heat exhaustion. *British Medical Journal* 10: 579-580.

72. O'Toole, M.L., P.S. Douglas, R.H. Laird, and D.B. Hiller. 1995. Fluid and electrolyte status in athletes receiving medical care at an ultradistance triathlon. *Clinical Journal of Sport Medicine* 5: 116-122.

73. Speedy, D.B., T.D. Noakes, I.R. Rogers, J.M. Thompson, R.G.D. Campbell, J.A. Kuttner, D.R. Boswell, S. Wright, and M. Hamlin. 1999. Hyponatremia in ultradistance triathletes. *Medicine and Science in Sports and Exercise* 22: 165-170.

74. Noakes, T.D. 1992. The hyponatremia of exercise. *International Journal of Sport Nutrition* 2: 205-228.

75. Speedy, D.B., T.D. Noakes, I.R. Rogers, I. Hellemans, N.E. Kimber, D.R. Boswell, R. Campbell, and J.A. Kuttner. 2000. A prospective study of exercise-associated hyponatremia in two ultradistance triathletes. *Clinical Journal of Sport Medicine* 10: 136-141.

76. Wolfson, A.B. 1995. Acute hyponatremia in ultra-endurance athletes. *American Journal of Emergency Medicine* 12: 441-444.

77. Irving, R.A., T.D. Noakes, R. Buck, R. Van Zyl Smit, E. Raine, J. Godlonton, and R.J. Norman. 1991. Evaluation of renal function and fluid homeostasis during recovery from exercise-induced hyponatremia. *Journal of Applied Physiology* 70: 342-348.

78. Rault, R.M. 1993. Case report: Hyponatremia associated with nonsteroidal anti-inflammatory drugs (abstract). *American Journal of Medical Science* 305: 318-320.

79. Noakes, T.D. 2000. Hyponatremia in distance athletes: Pulling the IV on the "dehydration myth." *The Physician and Sportsmedicine* 28: 71-76.

80. Convertino, V.A., L.E. Armstrong, E.F. Coyle, G.W. Mack, M.N. Sawka, L.C. Senay, and W.M. Sherman. 1996. American College of Sports Medicine position stand: Exercise and fluid replacement. *Medicine and Science in Sports and Exercise* 28 (1): i-vii

81. Casa, D.J., L.E. Armstrong, S.K. Hillman, S.J. Montain, R.V. Reiff, B.S.E. Rich, W.O. Roberts, and J.A. Stone. 2000. National Athletic Trainers' Association position statement: fluid replacement for athletes. *Journal of Athletic Training* 35: 212-224.

82. Basnyat, B., J. Sleggs, and M. Spinger. 2000. Seizures and delerium in a trekker: The consequences of excessive water drinking? *Wilderness and Environmental Medicine* 11: 69-70.

Chapter 9

1. Casa, D.J. 1999. Exercise in the heat I: Fundamentals of thermal physiology, performance implications, and dehydration. *Journal of Athletic Training* 34: 246-252.

2. Gleeson, M. 1998. Temperature regulation during exercise. *International Journal of Sports Medicine* 19: S96-S99.

3. Armstrong, L.E., Y. Epstein, J.E. Greenleaf, E.M. Haymes, R.W. Hubbard, W.O. Roberts, and P.D. Thompson. 1996. American College of Sports Medicine position stand: Heat and cold illnesses during distance running. *Medicine and Science in Sports and Exercise* 28 (12): i-x.

4. Casa, D.J., L.E. Armstrong, S.K. Hillman, S.J. Montain, R.V. Reiff, B.S.E. Rich, W.O. Roberts, and J.A. Stone. 2000. National Athletic Trainers' Association position statement: Fluid replacement for athletes. *Journal of Athletic Training* 35: 212-224.

5. Convertino, V.A., L.E. Armstrong, E.F. Coyle, G.W. Mack, M.N. Sawka, L.C. Senay, and W.M. Sherman. 1996. American College of Sports Medicine position stand: Exercise and fluid replacement. *Medicine and Science in Sports and Exercise* 28 (1): i-vii.

6. Casa, D.J. 1999. Exercise in the heat II: Critical concepts in rehydration, exertional heat illnesses, and maximizing athletic performance. *Journal of Athletic Training* 34: 253-262.

7. Holtzhausen, L.M., and T.D. Noakes. 1997. Collapsed ultraendurance athlete: Proposed mechanisms and an approach to management. *Clinical Journal of Sports Medicine* 7 (4): 247-251.

8. Noakes, T.D. 1998. Fluid and electrolyte disturbances in heat illness. *International Journal of Sports Medicine* 19: S146-S149.

9. Adolph, E.F. 1947. *Physiology of man in the desert.* New York: Interscience.

10. Gonzalez-Alonso, J. 1998. Separate and combined influences of dehydration and hyperthermia on cardiovascular responses to exercise. *International Journal of Sports Medicine* 19: S111-S114.

11. Morimoto, T., T. Itoh, and A. Takamata. 1998. Thermoregulation and body fluid in hot environment. *Progress in Brain Research* 115: 499-508.

12. Sawka, M.N., and E.F. Coyle. 1999. Influence of body water and blood volume on thermoregulation and exercise performance in the heat. *Exercise and Sport Science Reviews* 27: 167-218.

13. Sawka, M.N., W.A. Latzka, R.P. Matott, and S.J. Montain. 1998. Hydration effects on temperature regulation. *International Journal of Sports Medicine* 19: S108-S110.

14. Roberts, W.O. 1994. Assessing core temperature in collapsed athletes: What's the best method? *The Physician and Sportsmedicine* 22 (8): 49-55.

15. Armstrong, L.E., and C.M. Maresh. 1993. The exertional heat illnesses: A risk of athletic participation. *Medicine, Exercise, Nutrition, and Health* 2: 125-134.

16. Binkley, H.M., J. Beckett, D.J. Casa, D. Kleiner, and P. Plummer. 2002. National Athletic Trainers' Association position statement: Exertional heat illnesses. *Journal of Athletic Training* 37 (3): 329-343.

17. Dickinson, J.G. 1996. Predisposing factors, clinical features, treatment, and prevention. In *Hyperthermic and hypermetabolic disorders*, edited by P.M. Hopkins and F.R. Ellis, 20-41. New York: Cambridge University Press.

18. Eichner, E.R. 1998. Treatment of suspected heat illness. *International Journal of Sports Medicine* 19: S150-S153.

19. Epstein, Y. 2000. Exertional heatstroke: Lessons we tend to forget. *American Journal of Medicine and Sports* 2: 143-152.

20. Khogali, M., and M.K.Y. Mustafa. 1987. Clinical management of heat stroke patients. In *Heat stress: Physical exertion and environment*, edited by J.R.S. Hales and D.A.B. Richards, 499-511. New York: Excerpta Medica.

21. Knochel, J.P. 1996. Clinical complications of body fluid and electrolyte imbalance. In *Body fluid balance exercise and sport*, edited by E.R. Buskirk and S.M. Puhl, 297-317. New York: CRC Press

22. Armstrong, L.E., C.M. Maresh, A.E. Crago, R. Adams, and W.O. Roberts. 1994. Interpretation of aural temperatures during exercise, hyperthermia, and cooling therapy. *Medicine, Exercise, Nutrition, and Health* 3: 9-16.

23. Deschamps, A., R.D. Levy, M.G. Cosio, E.B. Marliss, and S. Magder. 1992. Tympanic temperature should not be used to assess exercise induced hyperthermia. *Clinical Journal of Sports Medicine* 2 (1): 27-32.

24. Knight, J.C., D.J. Casa, J.M. McClung, K.A. Caldwell, A.M. Gilmer, P.M. Meenen, and P.J. Goss. 2000. Assessing if two tympanic temperature instruments are valid indicators of core temperature in hyperthermic runners and does drying the ear canal help. *Journal of Athletic Training* 35 (2): S21.

25. Armstrong, L.E. 2000. *Performing in extreme environments*. Champaign, IL: Human Kinetics.

26. Brodeur V.B., S.R. Dennett, and L.S. Griffin. 1989. Exertional hyperthermia, ice baths, and emergency care at The Falmouth Road Race. *Journal Emergency Nursing* 15 (4): 304-312.

27. Clements, J.M., D.J. Casa, J.C. Knight, J.M. McClung, A.S. Blake, P.M. Meenen, A.M. Gilmer, and K.A. Caldwell. 2002. Ice-water immersion and cold-water immersion provide similar cooling rates in runners with exercise-induced hyperthermia. *Journal of Athletic Training* 37 (2): 146-150.

28. Costrini, A. 1990. Emergency treatment of exertional heatstroke and comparison of whole-body cooling techniques. *Medicine and Science in Sports and Exercise* 22: 15-18.

29. Hales, J.R.S. 1987. Proposed mechanisms underlying heat stroke. In *Heat stress: Physical exertion and environment*, edited by J.R.S. Hales and D.A.B. Richards, 85-102. New York: Excerpta Medica.

30. Hubbard, R.W., and L.E. Armstrong. 1988. The heat illnesses: Biochemical, ultrastructural, and fluid-electrolyte considerations. In *Human performance physiology and environmental medicine at terrestrial extremes*, edited by K.B. Pandolf, M.N. Sawka, and R.R. Gonzalez, 305-359. Indianapolis: Benchmark Press.

31. Hubbard, R.W., and L.E. Armstrong. 1989. Hyperthermia: New thoughts on an old problem. *The Physician and Sportsmedicine* 17 (6): 97-113.

32. Khogali, M. 1983. Heat stroke: An overview. In *Heat stroke and temperature regulation*, edited by M. Khogali and J.R.S. Hales, 1-12. New York: Academic Press.

33. Kielblock, A.J. 1987. Strategies for the prevention of heat disorders with particular reference to the efficacy of body cooling procedures. In *Heat stress: Physical exertion and environment*, edited by J.R.S. Hales and D.A.B. Richards, 489-497. New York: Excerpta Medica.

34. Knochel, J.P. 1996. Pathophysiology of heat stroke. In *Hyperthermic and hypermetabolic disorders*, edited by P.M. Hopkins and F.R. Ellis, 42-62. New York: Cambridge University Press.

35. Jessen, C. 1987. Thermoregulatory mechanisms in severe heat stress and exercise. In *Heat stress: Physical exertion and environment*, edited by J.R.S. Hales and D.A.B. Richards, 1-20. New York: Excerpta Medica.

36. Mayers, L.B., and T.D. Noakes. 2000. A guide to treating Ironman triathletes at the finish line. *The Physician and Sportsmedicine* 28 (8): 35-50.

37. Roberts, W.O. 1996. Environmental concerns. In *ACSM's handbook for the team physician*, edited by W.B. Kibler, 164-187. Baltimore: Williams and Wilkins.

38. Roberts, W.O. 1989. Exercise associated collapse in endurance events: A classification system. *The Physician and Sportsmedicine* 17 (5): 49-55.

39. Roberts, W.O. 1992. Managing heatstroke: On-site cooling. *The Physician and Sportsmedicine* 20 (5): 17-28.

40. Roberts, W.O. 1998. Medical management and administration manual for long distance road racing. In *IAAF medical manual for athletics and road racing competitions: A practical guide*, edited by C.H. Brown and B. Gudjonsson, 39-75. Monaco: International Amateur Athletic Federation Publications.

41. Roberts, W.O. 1998. Tub cooling for exertional heatstroke. *The Physician and Sportsmedicine* 26 (5): 111-112.

42. Seraj, M.E., A.B. Channa, S.S. Al Harti, F.M. Kahn, A. Zafrullah, and A.H. Samarkandi. 1991. Are heat stroke patients fluid depleted? Importance of monitoring central venous pressure as a simple guideline for fluid therapy. *Resuscitation* 21 (1): 33-39.

43. Shapiro, Y. 1987. Pathophysiology of hyperthermia and heat intolerance. In *Heat stress: Physical exertion and environment*, edited by J.R.S. Hales and D.A.B. Richards, 263-276. New York: Excerpta Medica.

44. Sutton, J.R. 1990. Clinical implications of fluid imbalance. In *Fluid homeostasis during exercise*, edited by C.V. Gisolfi and D.R. Lamb, 425-444. Carmel, IN: Brown and Benchmark.

45. Armstrong, L.E., R.W. Hubbard, W.J. Kraemer, J.P. Deluca, and E.L. Christensen. 1987. Signs and symptoms of heat exhaustion during strenuous exercise. *Annals of Sports Medicine* 3: 182-189.

46. Ayrus, J.C., J. Varon, and A.I. Arieff. 2000. Hyponatremia, cerebral edema, and noncardiogenic pulmonary edema in marathon runners. *Annals of Internal Medicine* 132 (9): 711-714.

47. Herfel, R., C.K. Stone, S.I. Koury, and J.J. Blake. 1998. Iatrogenic acute hyponatremia in a college athlete. *British Journal of Sports Medicine* 32 (3): 257-258.

48. Mann, S.O., and H.J. Carroll. 1992. Disorders of sodium metabolism: Hypernatremia and hyponatremia. *Critical Care Medicine* 20 (1): 94-103.

49. Mulloy, A.L., and R.J. Caruana. 1995. Hyponatremic emergencies. *Endocrine Emergencies* 79 (1): 155-168.

50. Shirreffs, S.M., and R.J. Maughan. 2000. Rehydration and recovery of fluid balance after exercise. *Exercise and Sport Science Reviews* 28 (1): 27-32.

51. Speedy, D.B., T.D. Noakes, I.R. Rogers, et al. 1999. Hyponatremia in ultra-distance triathletes. *Medicine and Science in Sports and Exercise* 31 (6): 809-815.

52. Speedy, D.B., I.R. Rogers, T.D. Noakes, et al. 2000. Diagnosis and prevention of hyponatremia in an ultradistance triathalon. *Clinical Journal of Sports Medicine* 10: 52-58.

53. Sterns, R.H. 1991. The management of hyponatremic emergencies. *Critical Care Clinics* 7 (1): 27-142.

54. Vrijens, D.M., and N.J. Rehrer. 1999. Sodium-free fluid ingestion decreases plasma sodium during exercise in the heat. *Journal of Applied Physiology* 86 (6): 1847-1851.

55. Armstrong, L.E., A.E. Crago, R. Adams, W.O. Roberts, and C.M. Maresh. 1996. Whole-body cooling of hyperthermic runners: Comparison of two field therapies. *American Journal of Emergency Medicine* 14: 355-358.

56. Armstrong, L.E., C.M. Maresh, S.A. Kavouras, D.J. Casa, J.A. Herrera-Soto, and F.T. Hacker. 1998. Urinary indices during dehydration, exercise, and rehydration. *International Journal of Sport Nutrition and Exercise Metabolism* 8: 345-355.

57. Montain, S.J., W.A. Latzka, and M.N. Sawka. 1999. Fluid replacement recommendations for training in hot weather. *Military Medicine* 164 (7): 502-508.

58. Murray, R. 2001. Regulation of fluid balance and temperature during exercise in the heat: Scientific and practical considerations. In *Exercise, nutrition, and environmental stress*. Vol. 1, edited by H. Nose, C.V. Gisolfi, and K. Imaizumi, 1-18. Traverse City, MI: Cooper Publishing Group.

59. Adner, M.M., J.J. Scarlet, J. Casey, W. Robison, and B.H. Jones. 1988. The Boston Marathon medical care team: Ten years of experience. *The Physician and Sportsmedicine* 16: 99-106.

60. Crouse, B., and K. Beattie. 1996. Marathon medical services: Strategies to reduce runner morbidity. *Medicine and Science in Sports Exercise* 28 (9): 1093-1096.

61. Elias, S., W.O. Roberts, and D.C. Thorson. 1991. Team sports in hot weather: Guidelines for modifying youth soccer. *The Physician and Sportsmedicine* 19 (5): 67-80.

62. Hargreaves, M., and M. Febbraio. 1998. Limits to exercise performance in the heat. *International Journal of Sports Medicine* 19: S115-S116.

63. Kleiner, D.M., and S.E. Glickman. 1994. Medical considerations and planning for short distance road races. *Journal of Athletic Training* 29: 145-151.

64. McCann, D.J., and W.C. Adams. 1997. Wet bulb globe temperature index and performance in competitive distance runners. *Medicine and Science in Sports and Exercise* 29 (7): 955-961.

65. Richard, C.R.B., and D.A.B. Richards. 1987. Medical management of fun-runs. In *Heat stress: Physical exertion and environment*, edited by J.R.S. Hales and D.A.B. Richards, 513-525. New York: Excerpta Medica.

66. Roberts, W.O. 2000. A twelve year profile of medical injury and illness for the Twin Cities Marathon. *Medicine and Science in Sports Exercise* 32 (9): 1549-1555.

67. Roberts, W.O. 1999. Mass participation events. In *Handbook of sports medicine.* 2d ed., edited by W.A. Lillegard, J.S. Butcher, and K.S. Rucker, 27-44. Boston: Butterworth-Heinemann.

Chapter 10

1. Rowell, L.B. 1986. *Human circulation: Regulation during physical stress.* New York: Oxford University Press.

2. Armstrong, L.E., and C.M. Maresh. 1991. The induction and decay of heat acclimatization in trained athletes. *Sports Medicine* 12: 302-312.

3. Armstrong, L.E., and C.M. Maresh. 1993. The exertional heat illnesses: A risk of athletic participation. *Medicine, Exercise, Nutrition and Health* 2: 125-134.

4. Armstrong, L.E., Y. Epstein, J.E. Greenleaf, E.M. Haymes, R.W. Hubbard, W.O. Roberts, and P.D. Thompson. 1996. American College of Sports Medicine position stand: Heat and cold illnesses during distance running. *Medicine and Science in Sports and Exercise* 28 (12): i-x.

5. Armstrong, L.E., and K.B. Pandolf. 1988. Physical training, cardiorespiratory physical fitness and exercise-heat tolerance. In *Human performance physiology and environmental medicine at terrestrial extremes*, edited by K.B. Pandolf, M.N. Sawka, and R.R. Gonzalez, 199-226. Indianapolis: Benchmark Press.

6. Armstrong, L.E., and C.M. Maresh. 1995. Exercise-heat tolerance of children and adolescents. *Pediatric Exercise Science* 3: 239-252.

7. Armstrong, L.E. 2000. *Performing in extreme environments.* Champaign, IL: Human Kinetics.

8. Convertino, V.A., L.E. Armstrong, E.F. Coyle, G.W. Mack, M.N. Sawka, L.C. Senay, and W.M. Sherman. 1996. American College of Sports Medicine position stand: Exercise and fluid replacement. *Medicine and Science in Sports and Exercise* 28 (1): i-vii.

9. Casa, D.J., L.E. Armstrong, S.K. Hillman, S.J. Montain, R.V. Reiff, B.S.E. Rich, W.O. Roberts, and J.A. Stone. 2000. National Athletic Trainers' Association

position stand: Fluid replacement for athletes. *Journal of Athletic Training* 35: 212-224.

10. Bergeron, M.F., L.E. Armstrong, and C.M. Maresh. 1995. Fluid and electrolyte losses during tennis in the heat. *Clinics in Sports Medicine* 14: 23-32.

11. Casa, D.J. 1999. Exercise in the heat. II. Critical concepts in rehydration, exertional heat illnesses, and maximizing athletic performance. *Journal of Athletic Training* 34: 253-262.

12. Rothstein, A., E.F. Adolph, and J.H. Wills. 1947. Voluntary dehydration. In *Physiology of man in the desert*, edited by E.F. Adolph, 254-270. New York: Interscience.

13. Armstrong, L.E., J.A. Herrera-Soto, F.T. Hacker, Jr., D.J. Casa, S.A. Kavouras, and C.M. Maresh. 1998. Urinary indices during dehydration, exercise, and rehydration. *International Journal of Sport Nutrition* 8: 345-355.

14. Montain, S.J., and E.F. Coyle. 1992. Influence of graded dehydration on hyperthermia and cardiovascular drift during exercise. *Journal of Applied Physiology* 73: 1340-1350.

15. Sutton, J.R. 1984. Heat illness. In *Sports medicine*, edited by R.H. Strauss, 307-322. Philadelphia: W.B. Saunders.

16. Hubbard, R.W. 1990. An introduction: The role of exercise in the etiology of heatstroke. *Medicine and Science in Sports and Exercise* 22: 2-5.

17. Knochel, J.P., and G. Reed. 1987. Disorders of heat regulation. In *Clinical disorders of fluid and electrolyte metabolism*, edited by C.R. Kleeman, M.H. Maxwell, and R.G. Narin, 1197-1232. New York: McGraw-Hill.

18. Speedy, D.B., T.D. Noakes, I.R. Rogers, J.M. Thompson, R.G.D. Campbell, J.A. Kuttner, D.R. Boswell, S. Wright, and M. Hamlin. 1999. Hyponatremia in ultradistance triathletes. *Medicine and Science in Sports and Exercise* 22: 165-170.

19. Montain, S.J., M.N. Sawka, and C.B. Wenger. 2001. Hyponatremia associated with exercise: Risk factors and pathogenesis. *Exercise and Sport Sciences Reviews* 29: 113-117.

20. Murray, R., W.P. Bartoli, D.E. Eddy, and M.K. Horn. 1997. Gastric emptying and plasma deuterium accumulation following ingestion of water and two carbohydrate-electrolyte beverages. *International Journal of Sport Nutrition* 7: 144-153.

21. Armstrong, L.E., and C.M. Maresh. 1996. Fluid replacement during exercise and recovery from exercise. In *Body fluid balance: Exercise and sport*, edited by E.R. Buskirk and S.M. Puhl, 259-282. New York: CRC Press.

22. Maughan, R.J., J.B. Leiper, and S.M. Shirreffs. 1996. Rehydration and recovery after exercise. *Sports Science Exchange* 9: 3.

23. Nadel, E.R., G.W. Mack, and H. Nose. 1990. Influence of fluid replacement beverages on body fluid homeostasis during exercise and recovery. In *Fluid homeostasis during exercise*, edited by C.V. Gisolfi and D.R. Lamb, 181-206. Carmel, IN: Brown and Benchmark.

24. Shirreffs, S.M., A.J. Taylor, J.B. Leiper, and R.J. Maughan. 1996. Post-exercise rehydration in man: Effects of volume consumed and drink sodium content. *Medicine and Science in Sports and Exercise* 28: 1260-1271.

25. Murray, R. 1987. The effects of consuming carbohydrate-electrolyte beverages on gastric emptying and fluid absorption during and following exercise. *Sports Medicine* 4: 322-351.

26. Casa, D.J., C.M. Maresh, L.E. Armstrong, S.A. Kavouras, J.A. Herrera-Soto, F.T. Hacker, N.R. Keith, and T.A. Elliott. 2000. Intravenous versus oral rehydration during a brief period: Responses to subsequent exercise in the heat. *Medicine and Science in Sports and Exercise* 32: 124-133.

27. Castellani, J.W., C.M. Maresh, L.E. Armstrong, R.W. Kenefick, D. Riebe, M.E. Echegaray, D.J. Casa, and V.D. Castracane. 1997. Intravenous versus oral rehydration: Effects of subsequent exercise in the heat. *Journal of Applied Physiology* 82: 799-806.

INDEX

A

acclimation or acclimatization (see heat acclimation/acclimatization)
acid-base balance 39
acquired factors in heat tolerance 166
acute heat neurasthenia (see transient heat fatigue)
acute renal failure 35-38, 50
adrenocorticotropic hormone (ACTH) 72
Advance Cardiac Life Support (ACLS) 189
air temperature (see dry bulb temperature)
air velocity 65
alanine aminotransferase (ALT) 32, 40, 59, 70
alcohol abuse 156,197
aldosterone 7, 79, 134, 141, 217-220
ambient temperature (see dry bulb temperature)
American College of Sports Medicine 134, 164, 188, 198, 201, 207-210
anaerobic metabolism 10
angiotensin II 8, 109
anhidrosis 138, 142, 154
anhidrotic heat exhaustion 20, 21, 23, 58, 140-141, 145
antidiuretic hormone (see arginine vasopressin)
arginine vasopressin 7, 111-112, 134, 164, 166, 217-220
aspartate aminotransferase (AST) 32, 40, 59, 70
athletes 51-56, 74-78, 91-95, 98-102, 114-119, 121, 152, 154-155, 157, 161, 166, 187-188, 193, 197-206, 209-210, 212
athletic trainer 52, 187, 197, 212
aural (ear canal) temperature 175
axillary temperature 42, 175

B

beta-endorphin 78, 79
black globe temperature 172-173
blood 38, 159
blood flow to skin (see cutaneous blood flow)
blood lactate 10
blood pressure 32, 33, 80, 86, 89, 158-160
blood urea nitrogen (BUN) 18
body temperature (see core body temperature)
body water (see total body water)
body weight 79, 105-107, 116, 125, 127, 199
brain edema 50

C

caffeine 156
capillary filling 69
carbohydrate-electrolyte fluid 98-100, 126, 191, 194, 204, 205, 208
cardiac arrest 108
cardiac arrhythmia 86, 87, 160
cardiac output 33, 80, 86, 89, 154

cardiovascular drift 11
cardiovascular strain 11, 13, 15, 33
cardiovascular system 1, 9, 11, 57, 75, 152, 217-220
central nervous system 1, 11, 32, 38, 43, 58, 75, 130, 170, 183, 217-220
central pontine myelinolysis (CPM) 130
cerebral edema 108, 133-134
children 14, 201
chyme 34
circadian rhythm 4
circulatory failure 78-80
classical heatstroke 31, 140
clothing (see uniforms)
coaches 170, 200
cold-sensitive neurons 3
collapse 21
coma 108, 130, 133
communication, medical coverage 192
conduction 1, 2, 153
convection 1, 2, 5, 153
cooling (see whole-body cooling)
core body temperature 4, 9-11, 31, 40, 68, 146, 154
creatine phosphokinase, creatine kinase (CPK) 18, 30, 32, 36, 37, 59, 70, 128, 157, 158
cutaneous blood flow 5, 12, 153, 155, 160, 175
cyanosis 62
cystic fibrosis 62, 164
cytokines 34

D

dehydration 7, 11, 31, 57, 66, 72, 77, 81, 91-92, 115, 117, 156, 157, 172-175, 182, 197, 201, 207-211
dermatologic conditions 138-149
dermis 145
dermoepidermal junction 144
deuterium oxide (D_2O) 72, 73
dew point 172
diarrhea 156-158, 201
dietary sodium 96-100, 118, 133-134, 141, 161, 205
discomfort index, heat stress 152
disseminated intravascular coagulation (DIC) 18, 30, 50
drinking behavior 77, 166
dry bulb temperature 65, 172-173, 203
dyspnea 62, 140

E

ear canal temperature (see aural temperature)
eccrine sweat glands (see sweat glands)
ectodermal dysplasia 154
edema 115, 141
education of athletes and competitors 174
electrolyte solution 71
electrolytes 30, 39, 110, 112, 114, 117-126, 202, 205
emergency medical technicians (EMT) 187

emergency plan 194
endocrine system 217-220
endotoxemia, endotoxin 34
environmental factors 166
epidermis 143, 147
epidemiology 39
equipment, care of patients 189-192, 209-210
erythema of skin 146
esophageal temperature 175
estrogen 4
etiology 60, 93
euvolemic hyponatremia 114, 115, 131
event directors 200
evaporation 1, 2, 5, 153
evaporative heat loss 2
exercise-associated collapse 60, 61, 68, 74-77, 82-86
exercise-heat stress 12, 13, 80, 154-155, 217
exercise-heat tolerance (see heat tolerance)
exercise performance 11, 31, 33, 77-80, 139, 154, 207
exertional heat cramps 20, 21, 24, 27, 91-102, 160, 180, 182, 209-210, 219
exertional heat exhaustion 21, 24, 27, 57, 57-90, 158-159, 173, 179, 209-210, 212-213,
      219
exertional heatstroke 5, 17, 18, 21, 23, 24, 26, 27, 29-56, 60, 61, 66, 68, 151-158, 172,
      173, 176-177, 209-210, 212-213, 218
exertional hyponatremia 20, 22, 23, 25, 57, 66, 68, 103-135, 162-166, 178, 182, 203,
      220
exhaustion 72, 108, 140
extracellular fluid 93, 97, 109-113, 115-122, 127, 130-131, 163-165

F
fever 5, 30, 154, 156, 157, 197, 201
fluid excess (see water overload, water intoxication)
fluid intake 120, 125-129, 204, 207-208, 211
fluid overload (see water overload, water intoxication)
follicular phase 4
folliculitis 137
football, the American sport 161, 174, 188

G
gastrointestinal tract 109, 158
glomerular filtration (see kidneys)
gram negative bacteria 34

H
headache 108, 131, 140
heart 38
heart rate 11, 33, 79, 154, 158, 169
heat acclimation/acclimatization 12, 59, 88, 100-101, 137-138, 148, 154, 158, 159,
      172, 197, 199
heat balance 1
heat cramps (see exertional heat cramps)
heat dissipation 32, 33
heat edema 20, 21, 141

heat exhaustion (see exertional heat exhaustion)
heat exhaustion II (see anhidrotic heat exhaustion)
heat gain 5
heat loss 5
heat rash (see miliaria rubra and prickly heat)
heat storage 1, 32, 33, 153
heat stress 12, 171-174
heatstroke (see exertional heat stroke)
heat syncope 20, 21, 22, 24, 27, 57-61, 68, 86-89, 159-160, 181, 182, 219
heat tolerance 138, 155, 158, 201
hematuria 35
hemoconcentration 8
hiker 126, 135, 163
history of heat illness 197
homeostasis 151
host factor 166
hot-wet environment 137, 141-144, 152, 159, 172
humidity (see relative humidity)
hydration status (see urine color) 193
hydrostatic force 8
hyperthermia 1, 5, 26, 31, 33, 40, 41, 43, 144, 151, 154, 156, 157, 171, 174
hypertonic fluid 158
hypertonic hyponatremia 131
hypervolemic hyponatremia 114, 115, 131
hypoglycemia 160
hypohydration 12
hyponatremia (see exertional hyponatremia)
hypotension 35, 160
hypothalamus 2, 3
hypotonic fluid 21, 107, 109, 158, 162, 203
hypotonic hyponatremia 113, 115, 131
hypovolemic hyponatremia 114, 115, 130-131

I

iatrogenic 130
ice water immersion (see whole-body immersion)
illness, bacterial or viral 27, 30, 34, 154, 156-158, 197, 201
immune system 34
infection (see illness)
inflammation 137, 142, 143, 146
inosine monophosphate (IMP) 11
incidence of illness 23-28, 65-67, 170-171
inherited characteristics as risk factors 197
International Amateur Athletic Federation (IAAF) 188
International Classification of Diseases (ICD) 19, 58, 74, 77
interstitial hydrostatic pressure 6
interstitial oncotic pressure 6
interstitial space 6
intracellular fluid 115, 122
intracorneal layer of skin 144
intraepidermal layer 143
intravenous fluid, IV fluid 42, 49, 66, 71, 72, 96-98, 132, 161, 190, 191, 205

K

keratin plug of sweat gland duct 142, 144
kidney failure 30
kidneys 7, 35, 38, 50, 105, 108-113, 132, 165, 217-220

L

laborers 139, 152, 154, 156, 161, 202-203
lactate dehydrogenase (LDH) 32, 40, 59, 70, 157, 158
lipopolysaccharide (LPS) 34, 35
liver 38, 50
lungs 38
luteinizing hormone (LH) 4

M

mammillaria 140, 144-145
marathon (see runner)
maximal aerobic power (see $\dot{V}O_2$max) 9, 13, 15, 118, 154
mean corpuscular hemoglobin concentration 118-121
Mecca 153
Medical aid station 188-191, 209-210
medical personnel and staff 53, 170-171, 187, 192, 193, 209-210
medications 153, 156
melanoma 146
mental performance 31, 139
mental status change 68, 70, 108, 131, 139
metabolic acidosis 66
metabolic heat production 1, 32, 153, 157
metabolism 2, 10, 11
mild heat fatigue (see transient heat fatigue)
miliaria crystallina 142, 144-145
miliaria profunda 142, 144
miliaria pustulosa 142-143
miliaria rubra 20, 21, 23, 137-138, 142-143, 154
minor heat illnesses 220
muscle cramps   63, 64, 91-102, 182, 191
muscle damage (see rhabdomyolysis) 126-128
muscle fasciculations 95
muscle glycogen 201
muscle potassium depletion 128
muscle twitches 95, 108
myoglobinuria 18
myxedema 5

N

National Athletic Trainers Association 166, 201, 211-213
nausea 108, 131
nephron 165
nonsteroidal anti-inflammatory drug (NSAID) 146-147, 165
normotonic hyponatremia 131

O

obesity 152, 155, 156, 197
older adults 15, 152-153

oral rehydration 72
oral temperature 42, 175
orthostatic hypotension (see postural hypotension) 71, 80
osmolality 6
osmotic 6
overdrinking (see water overload, water intoxication)
overhydration (see water overload, water intoxication)

P

patient history 70
photoallergic reaction 146
photosensitivity 146
physical performance (see exercise performance)
plasma oncotic pressure 6
plasma osmolality 131
plasma protein 6
plasma sodium (see serum sodium) 63, 64, 79, 104, 114, 118, 120, 122, 125, 127, 129, 132, 134, 164
plasma volume 6, 8, 72, 79, 112, 117-121, 127
platelet count 18
polyuria 140
postural hypotension (see orthostatic hypotension) 71, 80, 81, 159-160
potassium depletion 126-128
preoptic area, anterior hypothalamus 2, 3
predisposing factors for heat illnesses 151-167
prevention of exertional heat cramps 98-102
prevention of exertional heatstroke 44-45, 51-53
prickly heat (see miliaria rubra and heat rash)
progesterone 4
prognosis 39, 51
prothrombin clotting time 18
psychiatric disorders 87
pulse (see heart rate)
pulmonary edema 108, 133-134

R

radiation 1, 2, 5, 13, 153
recreational enthusiasts 197-206
rectal temperature 41, 42, 69, 70, 79, 154, 158, 169, 175, 183, 189, 199
rehydration, after exercise 204
relative humidity (see also hot-wet environment) 2, 32, 141, 169, 172, 175, 203
religious pilgrimage 153
rennin 7, 8, 109
respiratory exchange ratio 10
rhabdomyolysis 30, 35-39, 50, 126-128
road race 169, 209-210
runner (marathon, ultramarathon) 77, 103-106, 120, 125, 158, 163, 165, 188, 189, 209-210

S

saline (NaCl) solution 49, 96-98, 132-133
salt deficiency (salt depletion) heat exhaustion 20, 22, 57, 59, 62, 79, 81, 158, 161, 164
salt intake 7, 91, 95-102

scleroderma 154
seizure 87, 108, 126, 130, 133, 158, 191
serum sodium (see plasma sodium) 71-73, 95-97, 107-108, 113, 130, 164, 165
set-point temperature in the brain 2, 4
shivering 3
shock 80
signs and symptoms of exertional heat cramps 95-96, 200
signs and symptoms of exertional heat exhaustion 67, 200
signs and symptoms of exertional heatstroke 200
skin blood flow (see cutaneous blood flow)
skin cancer 146
skin disorders 138-149
skin rash (see also miliaria rubra) 140, 156
skin temperature 4, 8, 12, 14, 175
skin turgor 69, 81
sleep loss 27, 155-156, 197
sodium (salt) 7, 21, 22
sodium (salt) balance, whole-body 78, 81, 94-97, 100, 115, 117, 129, 162, 182
sodium (salt) deficiency 94, 96, 97, 100, 101, 160, 161-167, 182
sodium in diet (see dietary sodium)
soldiers 105, 133-135, 137, 152, 153, 157, 166, 188, 202-203
spinal cord injury 15
splanchnic organs 10, 33, 80, 109
splanchnic vasoconstriction 80
sports drinks (see carbohydrate-electrolyte fluid)
stable isotope 72
Starling forces 6
stratum corneum 144
stroke volume 11, 80, 154
subcorneal layer of skin 144
sunblock lotion 146
sunburn 20, 144-148, 154, 156, 197
supplies for care of patients 189-191, 209-210
sweat chloride (see sweat sodium)
sweating 3, 12, 13, 91, 92, 94, 97, 101-102, 105, 114, 119, 124-127, 137, 146, 156, 158,
    159, 202
sweat glands 138, 140, 142-146, 154, 217-220
sweat rate (see sweating)
sweat sodium 91, 94-97, 100-101, 126-127, 138, 141, 154, 163
syncope 10, 85-87

T

temperature regulation 2, 3, 9, 14-15, 146
tennis players 91-92, 101-102, 161
thermal strain 13, 171
thermoregulation 2, 3, 9, 29-56, 32, 146, 158
thermoregulatory failure 40
thirst 63, 64, 130
thyrotoxicosis 5
tolerance to heat (see heat tolerance)
total body water 6, 13, 72, 73, 122, 126, 163
total peripheral resistance 80
transient heat fatigue 20, 22, 138-139

treatment of anhidrotic heat exhaustion 149
treatment of exercise-associated collapse 84-85
treatment of exertional heat cramps 96-98, 185-186
treatment of exertional heat exhaustion 70, 185
treatment of exertional heatstroke 44-51, 184-185
treatment of exertional hyponatremia 185
treatment of heat edema 149
treatment of heat syncope 186
treatment of miliaria 145, 149
treatment of sunburn 149
treatment of transient heat fatigue 149
triathlete 105, 114, 116, 117, 125, 133, 164, 166
tropical anhidrotic asthena (see anhidrotic heat exhaustion)
tympanic temperature 42, 43, 175

U

Ultramarathon (see runner)
ultraviolet radiation from the sun 145-147
unacclimatized (see heat acclimatization) 197
uniforms 155, 157, 200, 202
Universal Precautions 189
urinary indices of hydration status 79
urinary sodium 63, 64, 154
urine 35
urine color 193
urine specific gravity 79, 195

V

vascular resistance 160
vasodilation 3, 160
vasopressin (see antidiuretic hormone, arginine vasopressin)
vasovagal syncope 85, 87
venous return 9, 10, 33, 89
VO_2max (see maximal aerobic power)
vomiting 63, 108, 131, 156-158, 201

W

warm-sensitive neurons 3
water balance 6, 115, 117, 129, 161-166
water deficiency heat exhaustion 20, 22, 59, 62, 81, 158
water overload, water intoxication 20, 23, 110-113, 114, 117, 121-128, 130, 135, 161-166, 206
weekend warriors 197-206
wet bulb temperature 172-173
wet bulb globe temperature (WBGT) 41, 152, 171-173, 177, 190, 193, 203
whole-body cooling 30, 41, 49, 169, 184-185, 189, 194, 202
women 13, 14, 163, 166

About the Editor

Lawrence E. Armstrong, PhD, fellow of the American College of Sports Medicine, is a professor in the department of kinesiology, Human Performance Laboratory, at the University of Connecticut. He is author of Human Kinetics' *Performing in Extreme Environments* (2000).

Dr. Armstrong received the Aerospace Medical Society's Environmental Science Award (1986), and the National Strength and Conditioning Association's Presidential Award for contributions to the NSCA *Journal of Environmental Physiology* (1989 and 1994). Since 1982, he has authored or coauthored 85 research articles for scientific journals and nearly 50 articles for educational and consumer publications. He also has contributed chapters to numerous books and government technical reports.

Dr. Armstrong also has personal experience with extreme environments. In addition to completing 14 marathons and climbing Mt. Washington four times, he has collected research data in the medical tent at the Boston Marathon. He has contributed to ACSM and NATA position stands on fluid replacement during exercise as well as position stands on heat and cold illnesses contracted during distance running. Dr. Armstrong graduated cum laude as a scholar-athlete from the University of Toledo in 1971 with a BEd in biology and comprehensive science, and he earned an MEd from Toledo in 1976 and a PhD from Ball State University in 1983 as a student of David Costill's. He is a former president of the New England chapter of the ACSM and conducted numerous research studies as a physiologist at the Research Institute of Environmental Medicine in Natick, Massachusetts, from 1983 to 1990. Dr. Armstrong lives in Coventry, Connecticut.

About the Contributors

Jeffrey M. Anderson, MD, is the director of sports medicine in the division of athletics at the University of Connecticut in Storrs, CT. Dr. Anderson has treated several heat-related illnesses among intercollegiate athletes (e.g., football, soccer, and track-and-field athletes) during his eight years in this position and also has served as the medical director for the Greater Hartford Triathlon. He has collaborated in several research studies conducted in the Human Performance Laboratory at the University of Connecticut.

Michael F. Bergeron, PhD, FACSM, is an assistant professor in the department of pediatrics at the Georgia Prevention Institute, Medical College of Georgia, in Augusta, GA. Professor Bergeron has worked with a number of junior, collegiate, and professional tennis players and other athletes to effectively prevent and treat heat-related muscle cramps. He also has conducted several studies on fluid–electrolyte losses and thermal strain during exercise–heat exposure and has gathered novel data related to individual heat cramp cases. Dr. Bergeron is the clinical applications editor of the *International Journal of Sport Nutrition and Exercise Metabolism.*

Douglas J. Casa, PhD, ATC, FACSM, is an assistant professor and director of athletic training education in the department of kinesiology at the University of Connecticut in Storrs, CT. Dr. Casa has served on two committees that wrote the NATA position statements "Fluid Replacement for Athletes" (chairperson) and "Exertional Heat Illnesses" (writing group member). Dr. Casa's previous research studies have evaluated medical protocols for rapidly cooling hyperthermic runners, urine color as an indicator of hydration status, influence of nutritional supplements (creatine, caffeine, glycerol) on heat tolerance and hydration status, and oral versus intravenous fluid replacement. He has served as an athletic trainer in the medical tents of the New York City Marathon and the Boston Marathon.

John W. Castellani, PhD, is a research physiologist at the U.S. Army Research Institute of Environmental Medicine in Natick, MA. He has conducted research studies and taught graduate courses at two New England universities concerning physiological mechanisms that limit an individual's ability to thermoregulate in hot and cold environments. He is an author or coauthor of more than 50 peer-reviewed journal papers, conference proceedings, and government technical reports. Dr. Castellani's practical experiences include assisting with the triage of heat exhaustion cases in the Boston Marathon medical tent.

Melissa P. Hazzard, MS, is a health fitness instructor at the Massachusetts Institute of Technology, in Boston, MA. She earned her BA in health and exercise science from Gettysburg College, in Gettysburg, PA, and recently earned her MS in exercise science from the University of New Hampshire, in Durham, NH. She has assisted with research examining rehydration methods during exercise in the heat, as well as the impact of hydration state on perceptual thirst responses in a cold environment.

Robert W. Kenefick, PhD, FACSM, is an associate professor at the University of New Hampshire, in Durham, NH. He has conducted research in the areas of hormonal control of fluid balance and metabolism and has published in numerous journals. In addition, he serves as a reviewer for scientific journals and is an editorial board member of the *Journal of Strength and Conditioning Research.* Dr. Kenefick has presented his research at numerous professional conferences, including the annual meetings of the ACSM and The Federation of American Societies for Experimental Biology.

Carl M. Maresh, PhD, FACSM, is director of the Human Performance Laboratory and head of the department of kinesiology at the University of Connecticut in Storrs, CT. As a former collegiate athlete, a collaborator on numerous studies involving fluid losses and performance in hot environments, and a consultant to many athletes during the past 20 years, Dr. Maresh brings particularly interesting insights to this book. Dr. Maresh has written and coauthored more than 100 scientific publications, including investigations conducted in the areas of fluid–electrolyte balance, hormonal responses to exercise and environmental stress, and sport nutrition. He is a past president of the New England Chapter of the ACSM and member of the ACSM Board of Trustees.

William O. Roberts, MD, MS, FACSM, is in private practice with specialty boards in family practice and sports medicine with MinnHealth Family Physicians in White Bear Lake, MN. Dr. Roberts has served as a leader in ACSM. He is a senior associate editor and a regular contributor to *The Physician and Sportsmedicine.* Dr. Roberts has published several articles and chapters on recognition and treatment of exertional heatstroke, other exertional heat illnesses, and body temperature measurement in athletes. He is the medical director of the Twin Cities Marathon and on the medical advisory boards of the USA Cup Soccer Tournament and the Minnesota State High School League.

Jaci L. VanHeest, PhD, is an assistant professor at the University of Connecticut in Storrs, CT. She formerly held the position of director of physiology for USA Swimming in Colorado Springs, CO. Dr. VanHeest worked intimately with the highly successful U.S.A. swim team that garnered many gold medals at the 1996 Summer Olympics in Atlanta. As a former collegiate athlete herself, Jaci understands athletes well and has been successful in helping athletes to optimize their performance and health. Her publications include research in the areas of exercise metabolism in females, body composition, sport nutrition, and human performance in competitive athletes.